Nanotechnology Revolution

By

Chakrapani Srinivasa

Nanotechnology Revolution

By

Chakrapani Srinivasa

Copyright 2021 Chakrapani Srinivasa

About the Author

Chakrapani Srinivasa (Padmaja), Freelance journalist from India possesses Bachelor degree in Engineering (B.E) and Post graduate in Business Management (MBA) with Distinction. He has worked as Associate Editor of 'Naradar' fortnightly journal in Chennai, India. He is the Senior Editor of the journal "The Divineness".

Contributed articles, short stories and travelogues in leading journals like Ananda Vikatan, Kumudam, Savi, Kalki, Dinamani Kadhir, Dinamani daily, Idhayam Pesukirathu, Naradar etc

He has written articles and e books through many international publishers all over the world.

He is the Consulting Editor: Contemporary Who's Who-Research Board of Advisers of ABI, USA.

Preface

Nano technology has created a revolutionary scenario in India.

Many clamor to know about it, as it is yet to reach the common man and the public. The government's efforts to enlighten the young and budding entrepreneurs to take up manufacturing efforts are the order of the day. Only MNCs and Research Institutes with sound financial background and sophisticated R&D set up, take a plunge in it. Others hesitate as the common man in the streets will not respond to it and the investment will lead to loss commercially. Topics covered in this book are:

-Nano Composites and Manufacturing Processes,

-Quantity and raw materials used for producing nano composites materials,

-Nano Technology in Satellites, R&D undertaken in Nanotechnology for Satellites,

-Indian Health Care and Nanotechnology

-Purpose, system and materials used for production of nano composites

-Companies, Institutes and Scientists involved in Nano Composites

-Scientists and Institutes participated in Bangalore India Nano 2016

-Exhibitors in India Nano 2016

-Trade Fairs attended by Nano Composite Scientists and Subjects covered

-Nanotechnology for agriculture

-Current research projects in India

Contents

Nano Composites and Manufacturing Processes........6

Quantity and Raw Materials Used for Producing

Nano Composites Materials..9

Manufacturing Processes..13

Nano Technology in Satellites......................................16

R&D undertaken in Nanotechnology for Satellites....................20

Indian Health Care and Nanotechnology....................26

Purpose, System and Materials Used for Production Of Nano Composites....................40

Scientists and Institutes participated in Bangalore India Nano 2016....................44

Exhibitors in India Nano 2016....................................50

Trade Fairs attended by Nano Composite Scientists and Subjects covered....................55

Companies, Institutes and Scientists involved
in Nano Composites……………………………………..57

Nanotechnology for Agriculture........................64

Current research projects in India......................74

Nano Composites and Manufacturing Processes

Indian experts use fiberglass and reinforced plastics for multifarious applications. Still research is done by them to achieve innovation in Nano composites with attractive properties for defense needs.

Knowledge on Nano Technology was made aware in India only 2001.

We are far behind countries like Japan, USA, Europe, and Taiwan in the field of research based on Nano Composite to enhance the material properties.

The Indian scientists have shown interest to manage the Nano structures by new synthetic methods. They have realized that properties of Nano composite materials are based not only on the properties of their individual parent materials but also on their morphology and interfacial nature. Utilizing suitable fabrication processes and controlled Nano-sized second phased dispersion, they were capable of improving the toughness, flexural strength, thermal stability and obviously possess betterment in Nano-dispersion.

At low cost they could produce Nano composite materials to be used for medical field, automobile parts, electronic goods and defense. They have Nano fillers dispersed in the matrix.

The Nano- composite materials ranges as follows:

1) 3 dimensional metal matrix
2) Dimensional lamellar composites
4) Nano wires of single dimension.
5) Zero dimensional core-shell.

Indian experts use fiberglass and reinforced plastics for multifarious applications. Still research is done by them to achieve innovation in Nano composites with attractive properties for defense needs.

The structure of Nano composites has matrix filler combination, where the fitters surround and binds together to form discrete units in the matrix.

The Indian industrialists and entrepreneurs engage fiber glass and reinforced plastics as composite materials for their production processes.

The Nano materials have totally different physical, chemical and biological properties compared to the properties of individual atoms and molecules. The Nano particles have the special feature to alter their magnetic nature, melting temperature, color etc without altering the material's chemical composition. They are made fit for usage in polymeric materials since they have very high surface to volume and aspect ratios. This structure improves the super conducting and material properties in many useful applications for innovations. The properties of Nano composites will be such that they are extremely tougher than the bulk component. Indian research scientists are involved in Nano composite organic and inorganic materials.

The inorganic components like Zeolites, and 3 dimensional inorganic components clays, metal oxides Metal phosphates are 2 dimensional layered materials. Mo3Se3 and clusters form the one dimensional and zero dimensional Nano composite materials. They find a place in Nano wires, battery cathodes, light weight components sensors and non-linear optics. These inorganic layered materials have a good defined and ordered intramellar space, which can be accessed by foreign species. This phenomenon supports to act as matrices for polymers to produce hybrid Nano-composites.

In lamellar Nano composites, interface interactions between two phases are high. With polymer host interactions we can achieve Nano composites with various properties for individual growth and defense equipments.

Lamellar Nano composites have 2 versions as follows:

1. Intercalated.
2. Exfoliated.

In intercalated Nano composites we have polymer chains alternately present with inorganic layers in a fixed compositional ratio. They will acquire a well defined number of polymer layers in the intramellar space. But in the case of exfoliated Nano composites, the number of polymer chains between the layers is continuously variable. It is found that layers are > 100 A apart.

The intercalated Nano composites find place in electronic and charge transport properties.
Very high and better mechanical properties like dielectric constants, which are flexible and easy to process are seen in exfoliated Nano composites.

Still the scientists are doing research to locate a Nano composite material to have all these properties.

Quantity and Raw Materials Used for Producing Nano Composites

Materials

Polymer materials are easy to process and possess low dielectric constant compared to ceramic materials used in many Indian industries, which are brittle and processed at high temperature

Indian industrialists are engaging materials with high dielectric constants with flexibility and easy to process.

The composite materials, which have micron scale ferroelectric ceramic particles as the filler in liquid crystal polymer or thermoplastic polymer matrices do not have the required processing phenomenon. Hence they are not capable enough to form into thin uniform films. So the importance of Nano composites materials with many materials mixed at nanometre scale is highlighted for various applications in industrial developmental activities.

Indian market has many commercial Nano particles that can be formed into polymer matrix to produce polymer Nano composites.

The scientists will choose the right particles to achieve the required penetration of the polymer or its precursor into the interlayer spacing of the reinforcement. This will produce exfoliated or intercalated product as needed.

The specialty of polymer nano composites is that it comprises of polymeric material like thermo sets and thermoplastics elestomers along with reinforcement of nano particles

The Nano experts say that polymer will be incorporated as polymeric species or via the monomer, which is polymerized to produce corresponding polymer clay nano composites.

The Nano particles, which are preferred by experts, are:

1. Carbon Nano fibbers

2. Nano Silica

3. Polyhedral Oligomeric Silsesquioxane

4. Carbon Nano tubes

5. Nano Aluminium Oxide

6. Nano Titanium Oxide

7. Montmorillomite Organo Clays

Following Thermosets and Thermoplastics are utilized as matrices for producing Nano composites

1) Epoxy resins

2) Polyimides

3) Polystyrene

4) Polyethylene Terephthalate

5) Poly Urethanes

6) Polyo lefin

7) Nylons

During the process of manufacturing Nano composite materials we face 2 types of obstacles after the selection of the desired nano particles for the requirement. They are:

-The interfacial interaction and or compatibility with the polymer matrix are needed for the selection of nano particles.

-A good uniform dispersion and distribution of nano particles or nano particle aggregates within polymer matrix is a must for the processing method followed.

The mechanical properties are fully dependent upon the quantity of nano particles later added to polymer matrix, which will play an important role for acquirement of the mechanical properties by the nano composites.

The scientists in the production unit add these in small quantities to get better properties. This process will fetch reduction in weight as desired by the defense men and aircraft crew. It also adds more strength, sturdiness and barrier capabilities. To get the same performance micro dimensional particles will need higher loading levels than this.

The researchers in this field have expressed that Nano clay modification of polymers like Polyamides may lower the impact performance. This is considered as a disadvantage while utilizing the Nano particles.

The scientists are happy to get the following advantages through Nano particles:
1. Gas barrier
2. Thermal expansion
3. Dimensional stability
4. Synergistic flame retardant additive
5. Tensile strength toughest and stiffness
6. Chemical resistance
7. Reinforcement
8. Ablation resistance
9. Thermal conductivity

The disadvantages regarding the nano particles, which worry the scientists, are
1. Sedimentation
2. Dispersion setbacks
3. Black colour when various carbon containing Nano particles are utilized
4. Optical issues
5. Increase in viscosity

Manufacturing Processes

It is important to know that the polymer nano composites use fillers having dimensions on a nano meter scale reinforced into the polymer matrix. These materials add Nano filler with a polymers to give a composite, which are equal or with better physical and mechanical properties. Compared to their conventionally filled counter parts it will have lower loadings of fillers.

Since we have the increased surface area with Nano fillers, the polymer Nano composites fetch the potential for better mechanical properties along with thermal strength barrier and flame retardant capabilities than their conventionally filled materials.

The system and process followed at present for getting nano composites is taking individual steps for polymerizing each of the various monomers and then followed by pelletization of each of the various polymers thus formed separately.

Then these individual pellets will be mixed with a Nano filler material in an extruder to form the nano composites material.

This process is considered as a costly affair even though it is regarded by experts as an efficient way to form nano composites.

Types of nanomixed and layered materials are as follows:
-3dimensional metal matrix composites
-2dimensional lamellar composites
-One dimensional nano wires
-Zero dimensional shells

These types of construction combinations produce excellent properties of each of the individual components.

They contribute totally new properties to be used for defense medical and environment. Another process to produce nano composites is by multilayered silicate material into a thermoplastic polymer at a temperature higher than the melting or softening point of the thermo plastic polymers.

This polymer is a chosen from the category having of a thermoplastic urethane, thermoplastic epoxy, polycarbonate, polyester, nylon etc and their mixtures.

Another interesting process is that the Carbon nano tube-reinforced composites can be synthesized utilizing a powder mixing process with a powder-powder mixing between carbon nano tubes and ceramic powder or raw metal like aluminium or copper matrix. Then the process of conventional sintering is done.

The Indian experts have found that the behaviour of these Carbon nano tubes reinforced composite materials have shown a lower percentage in mechanical properties. It is found that relative density of the materials after sintering process is less .This is obviously revealed that many fracture sources like defects and pores could result in low mechanical properties. The main reason for this is due to intensive agglomeration of carbon nano tubes on the metal powder surface and the use of conventional consolidation processes

Anyhow the nano tubes agglomeration in a metal matrix can be avoided by homogeneous dispersion of Carbon nano tubes in the metal matrix.

The Carbon nano tubes can be dispersed in a pre determined dispersing solvent like Nitric acid solution, Ethanol, Water, N-Dimethyl formamide, Thionyl chloride and Dichlorocarbene to produce a dispersed solution, which is then treated further using ultrasonic waves. With ultrasonic waves treated dispersed solution, water soluble metal salts or metal hydrates are blended and dried to eliminate water vapour, nitrogen and hydrogen. Then at last stage it is calcinated to get a stable carbon nano tubes nano composite powder.

For high valued abrasive materials or wear resistant coating materials we can consider the metal nano composites as the best choice for aerospace, medical and for machine parts, which need the metal nano composites powder. Its capability to become bulky at faster rate is added advantage. High quality sintering performance of it compared to conventional metal composite materials is a special feature for usage in industrial or defense applications. Whenever lighter weight is need as in defense equipments, the lower density polymeric materials are the best by including thermally and electrically conducting fillers. It develops conducting type nano composites.

Automobile industry uses the Thermoplastic nano composites for it's under hood applications and interior parts. The plastic wrap and for carbonated beverage containers this is find useful.

Thermoplastic and thermosetting have the differences in process cycle, structural components, recyclable reprocessing nature, durational of the process cycle etc.

Thermoplastic resins can be used for toys, domestic applications, automobiles etc but thermosetting resins can be used for marine aerospace industry and defense equipment

To produce thermostat nano composites integration of nano particles with polymers is done these are complex hybrid materials with quite different properties and characteristics for various industrial production activities the researchers do not so much care for thermostat nano composites compared to thermoplastic nano composites at present some have taken steps to produce it with updating of knowledge on characteristics of the inter phase –region and the calculation of technology –structure property relationships many nano particles are used to prepare nano composites with thermosetting polymers along with diamond alumina graphite and smectite clay these thermostat nano composite gives better thermal stability chemical resistance flame retarding capabilities . Hence marine industries construction field etc engage them to fetch better customer satisfaction they have the advantage of shaping with injection moulding and extrusion processes.

For packing and fibre industry these materials are useful.

Nano Technology in Satellites

Usage of Nanotechnology in satellites is an attempt to minimize the size and cost of components assembled in it.

This is a new field and for countries like India, the scientists need lots of exposure through foreign experts.

Nanotechnology is rightly termed as a gentle introduction to the next big idea. For better imagery satellites and improved solar energy technology, sophisticated space solar cells are a must.

Nanotechnology supports to improve solar cells and performance of satellites. Indian scientists strive to the next level of developing nano structured materials and subsequently Nano Structured Cells. This will lead to better usage of solar energy by the scientists.

With the team of National Research Professors from Indian Institute of Science in Bangalore, Karnataka's Vision Group on Nanotechnology, Department of Science and Technology, Government of India, United Nano Tech Innovations P Ltd in Bangalore, Chemistry and Physics of Materials Unit, JNCASR in Bangalore, IISER in Pune etc they analyze the fundamental behaviour of Nano Scale Crystals, which change the manner in which solar cell absorbs light and converts it into electricity needed for developing country like India.

This power stricken nation needs Nano Technology for betterment of the existing solar cell technology, which has 3 individual Photovoltaic junctions used in a series. Germanium, Gallium Arsenide and Indium Gallium Phosphide form the triple junction solar cells.

The Indian scientists are involved in augmentation of the middle cell with a quantum dot array for betterment of short circuit current and enhance the total efficiency of the cell.

Their research work is supported by Advanced Materials & Green Chemistry, Tata Chemical Ltd in Pune, National University in Singapore, Bio-materials National Institute of Nano Technology in Canada, Materials Science & Engineering in Stanford University, CA, USA and Nanotechnology Initiative Council in Iran.

To achieve better performance in propulsion systems Nanotechnology contributes a lot. It improves high performance materials with Nano sensors. Though the present rocket engines operate on chemical propulsion system, their attempts to get new forms of space propulsion systems like EP system, is an encouraging factor.

Electric propulsion with Field Emission Electric Propulsion, Colloid Thrusters etc. reduce the propellant mass and permit to enhance pay load capacity or reduce the launch mass. Indian scientists need the assistance of Japan and USA which have successfully implemented in their space mission like Hayabusa, ESA's smart – 1.

The Electric Propellant system use the electro statically charged and accelerated Nano particles as propellant. Millions of micron sized Nano particle thrusters will be fitted in 1 sqcm for a good fabrication of highly scalable thruster arrays.

Nano technology effectively shields from radiation effects. To achieve less weight, structural stability and radiation protection we need to erect multifunctional space craft hulls. Sophisticated Nano materials like isotopic ally enriched boron nano tube could aid to get Nano sensor – integrated hulls to have shielding from radiation and better energy storage capability.

The onboard electronics should also be protected and Nano technology supports it well. The electronic gadgets in satellites are more tolerant to radiation if their size is reduced. It has been found that Quantum dot devices are tolerant to radiation hundred times more than the bulky conventional devices.

Quantum dot / CNT based photovoltaic devices have proved to be highly resistant compared to the age old heavy and bulky solar cells.

India with great financial difficulties is managing satellite projects. Hence protection of their invaluable satellites in space is highly essential. The issues created by ASAT missile systems, the DRDO and ISRO scientists in India are eager to save their precious satellites with quantum dots. These quantum dots will emit radiation profile identical to that of their satellites.

India is yet to learn a lot from scientists and Research Institutes in USA, as the neighbouring countries like China and Russia have developed Ground Based Directed Energy Weapons to cause threats to satellites. Calamity through high powered micro waves is another great concern for Indian Satellite scientist. They have to in depth in Nano technology studies to gain efficiency to manage any space threats.

Nano materials are another area in which Indian Satellite experts show interest. With blacker than pitch nano materials we can collect scientific measurements which are tough to obtain. It will also aid to view some astronomical objects which are not visible now. This nano material is a thin coating of multiwall carbon nano tubes. This will prevent the damages occurring to components in satellites from errant light.

Indians are yet learn the skill to develop Nano fabrication, precision engineering skill, sophisticated lithography to design space instruments with improved and safe performance. High resolution x – ray spectrometers, solar physics instrumentation etc. is some of their needs in their space research projects. They have to be trained well in institutes like MIL and NASA to specialize in it.

Nano electrical mechanical system focuses on miniaturization which is the next improved stage from micro electro mechanical systems.

They function on Nano scale rather than micro scale. They possess lesser mass and greater mechanical resonance frequencies, zero point motion and greater surface to volume ratio. These are very useful for surface based sensing mechanisms.

They create physical, chemical and biological sensors to detect chemical substances in the air. As they are replacing the MEMS, the Indian Research Organizations like Indian Institute of Science in Bangalore, Indian Institute of Technology in New Delhi, Kharaghpur etc. gain support from internationally reputed institutes like National Institute of Nano technology in Canada, materials science & Engineering, Stanford university, CA in USA, Ecole Normale superrieure in Paris, etc.

The Indian Satellites are gained by reduced size, lower power consumption and economical investment with NEMS.

The main application of NEMs is Atomic Force Microscope tips. As the NEMs have higher sensitivity we can produce valuable sensors in satellites to identify stresses, Chemical signals and vibrations. The special feature in it is the carbon Nano tubes of small diameter and high aspect ratio mounted to the tip to get perfect resolution of the micro scope. They have 1nm to 50nm diameter single wall and multiwall carbon Nano tubes, which can be used for wires in Nano technological devices in satellites.

To meet needs and features of NEMs, materials used in it are carbon Nanotubes, graph ere and diamond which are carbon based. This carbon enables stability and support to function as transistor. This is suitable for satellites in orbit as they have low density, low mechanical dissipation and low friction with large surface area. Switches, Nano motors and high frequency oscillators are the contribution of NEMs in satellites, Due to existence of carbon Nano tubes and Grapheme, ability to manage higher stress is possible in NEMs devices used in space.

NEMs technology in India has paved way to do research in Nano accelerometers, integrated piezo resistive detection devices etc. in many IITS and institutes like SASTRA, IISER in Pune. Valuable applications achieved through their research studies are Energy harvesting, accurate sensing, displaying and imaging etc.

NEMs sensors, Nano electrometers, relays with Nano tubes, NEMs memory have found place in Indian Satellite system study.

Developed countries like Germany, USA, and Japan take a leading role in R & D activities in Nano technology for satellites. Indian scientists have to positively get updated with their assistance.

R&D undertaken in Nanotechnology for Satellites

Main aim of Nanotechnology is to make space projects to be successful and practical with light weight, economical and better performance.

It also eliminates the delays linked with space based services. Reduction in Space Craft Life Cycle Costs and Lead Time to take Space Based Systems to be competitive with Ground Based Systems are the basic features of all R & D works undertaken with Nano Technology. Large volumes of data with accuracy in shorter time span can be obtained by this technology. Also it has the advantages of minimum direct control and maintenance.

The R & D works in leading Institutes like IIT (Delhi, Mumbai, and Ahmadabad), SRM, University of Rajasthan, Amity Institute of Nanotechnology, Alagappa University, and SASTRA in Tanjore etc, involve in bottom up techniques, which will structure the material from the atomic and molecular level. Controlled Nano-structuring is their aim. Compared to bulk materials, the Nano materials have different properties to face any issues in satellites.

The researchers work for advancements in Nano materials to obtain Light Weight Solar Sails and Cable for Space Elevator. The need for Rocket Fuel is also brought down along with time and cost to reach the orbit and travelling in space. With this Nanotechnology we can produce Nano Sensors and Nano Robots to enhance the performance of satellites.

Their intensive study in Nanotechnology has given light to Carbon Nano Tubes to lower the weight and also increase the strength of the structure. By deploying a network of Nano Sensors, more data regarding Mars and water availability in it could be obtained.

To minimize the complications in Thruster Systems, the MEMS devices are used to accelerate Nano particles.

To monitor the levels of trace chemicals in space craft and to enhance the performance of Life Support Systems, attempts are taken by the Indian Scientists in various research institutes.

Of course they need the guidance, training and technical knowhow from Centre for Nano Technology at NASA Ames, Johnson Space Centre Nano Materials Project, Lift Port Group and Space Nano Technology Lab at MIT.

The Indian Scientists focus their attention on fundamental behaviour of Nano Scale Crystals (Quantum Dots), which change the manner in which a solar cell absorbs light and produce electricity out of it. By altering the size of the particle, the electrical, mechanical, optical and thermal properties of Nano materials can be changed and controlled.

These new characteristics will be an invaluable contribution to development of Semiconductor Devices in satellites. All information related to surveillance, weather, navigation, missile warning and communication capabilities can be obtained from satellites in the orbit.

Indian military and the common man are fully dependent upon satellite systems for TV, Mobile Phone services, R & D activities, Mobile Broadband Services, GPS etc. Any damage or threat to satellites will doom the society and cripple their vital activities. At this juncture R & D works on Nano Technology to overcome such threats is essential now for our country.

Protection from Ground Based Directed Energy Weapons to Power System Components, Electro Optical Sensors, Thermal Control and Structure of satellites is the greatest concern for the Indian researchers on Nano Technology. Low Power Lasers will destroy or jam Satellite Electro-Optical Sensors and High Power Lasers will damage by overheating parts of the satellite. Both in-band damage and out of band damage can be solved by Nano Technology.

Nano structured surface coating will harden the satellite surface to reflect, absorb or transmit the incident energy. It may even perform all the three operations together to save the satellite. Dispersion of thermal and electrical energy across the surface will also protect the satellites from calamity. This is done by utilizing Carbon Nano Tube Membranes or Bucky Papers to render adequate Electromagnetic Shielding and improve Lateral Thermal Conductivity.

Research on shielding with Carbon Nano Tube forests, which will emit electrons to the atmosphere and generate Plasma Shield around satellite structure is done in India.

Indian scientists should be enlightened by the R & D works in AFRL, University of Dayton to tailor the electrical conductivity of Polymer materials, which are utilized to build Commercial and Military Aerospace components.

If Carbon Nano Tubes with 50nm to 150nm diameter and possessing current capability of copper are carefully dispersed into a Polymer Matrix, then it will produce electrically conductive Polymer Composite Nano Material Based Shield from radiation.

Research on protection from electromagnetic interference using Nickel Nano strands, which have electromagnetic, metallurgical and chemical properties, is being done. Still more intensive research is to be carried out to protect satellites from Ground Based Directed Energy Weapons using Nano Technology.

India has to prepare well to tackle the neighbourhood attacks on their satellites. Already India is in darkness with rising population, illiteracy, language and caste issues. Our country should not suffer even from day to day activities by killing the satellite links. So, R & D in Nanotechnology is a must.

With the guidance and research assistance from NASA Goddard Space Flight in Greenbelt, Md, the Indian Scientist in satellite study have to be enlightened about Nano material to suppress errant light that has a peculiar way of bouncing off instrument components and destroy all measurements.

Intricacies in Nanotechnology covering Nano fabrication, advanced Lithography and accurate engineering technology to erect sophisticated instruments, Solar Physics Instrumentations are needed for Indian Scientists with the assistance of MIT, Space Nano Technology Lab.

Tech Comp India P ltd in New Delhi is engaged in Bio Safety lab for the past 7 years, as per the norms stated in International manual on Bio safety in Micro Biological and Bio Medical labs level 1-5NIH/CDC USA, and Bio safety level manual 3rd edition Bio Safety level 1-4 WHO Geneva .

They have the special features to prevent the exposure to Swine Flu, TB, and Anthrax etc, which killed many in India.

They offer designs as per the requirements like Prefabricated, Lab on Wheel suitable to operate in villages in India. The staff employed are trained abroad to maintain standard and quality.

Contact address: Tech comp India P Ltd 906, 9TH Floor, Pearls Best Heights,-1, Plot A5, Netaji Subhash Place, Pitampura, New Delhi, 110034, Phone: 08042985715.

With a turnover of Rs 500 crores they engage 90 staff to manufacture all lab equipments, Blood bank Centrifuge, Deep Freezers, Transfusion table, High Speed Centrifuge, Incubator, Stability Chamber, Refrigerated Centrifuge etc.

They have business contacts with Japan, Korea, Singapore and USA,

Agaram Industries is located in 73, Nelson Road, Amjikarai, Chennai, 600029. Phone: +91-44-23741413, fax: 91-44-23741529. Email:sales@agaramindia.com.

This company deals with imported Pasta making machines, Pasta Dryers and Pasta Lines manufactured by Imperia

Monferrina in Italy. Mixolab, NIR Analyzer, Alevaolab and Mixolab from Chopin France is also supplied by them. Products of Bastak in Turkey are marketed by them in India. Gluten Analyzer, Silo and Grain handling equipments are their popular sales products in India.

Their marketing team handles the online Refractometer, TOC Analyzer imported from 01 Analytical USA; Tensiometers and Contact Angle Meters imported from Kyowa Interface Japan; Magnetic Stirrers, Hotplate Stirrers and Melting Point Apparatus imported from Bibby Scientific U.K. Digital Saccharimeters, Digital Density Meters, Digital Polarimeters imported from Rudolph Research Analytical USA. This reputed company also market local India made Friability Tester, Hardness Tester and Bottle Cap Torque Tester with good after sales service team trained abroad.

Liquid Crystal Thermal Sensors for Thermal Sensing Application was vividly displayed by Centre for Nano and Soft Matter Science.

Contact person: Dr.C.V .Velamaggad, a Scientist in this Centre. This is a cost effective commercial strips, which can be used for personal use of a common man.

Silicon AFM probes from Budget Sensors located in Innovative Solutions Bulgaria Ltd. 48, Joliot Curie Str, 1113 Sofia, Bulgaria, drew the attention of many delegates. Single Hi-Res AFM Probes, Magnetic AFM Probes, Conductive AFM Probes and Silicon AFM Probes are their products. Single Wall Carbon Nanotubes introduced by OCSiAI has been successful in battery applications. Aluminium coated with single wall carbon Nanotubes reduces the interfacial resistance,

The conference and exhibition enabled many scientists to exhibit their inventions. They are as follows:

-Metallic nano brushes, which can capture micronized objects like bacteria, is an invention of a scientist Depanjan Sarkar, Unit of Nano Science, IIT Chennai
-Mobile Phone Based Next Generation Plasmonics - developed by Venkatesh.S from Sri Sathya Sai Institute of Higher Learning, Puttaparthi, Ananathapur, A.P
-Fast Responsive Soft Biometric Actuators, an innovative idea by Rohit Goyal, Centre for Nano and Soft Matter Sciences,

Low cost ZnO Nano particles based OFETs for Co_2 gas detection by Ashwath Narayana.B.S, from Visvesvaraya Technological University.

-Application of Aluminium Oxide Nano particles and TiO_2 Nano particles to treat heavy metals – K.R. Sreeharsha, VTU CPGSB, Muddenahalli, Chikkaballarpur,

-Technology Development of Nanostructures – Titania Microspheres for Self Cleaning Textiles application – Yamini Ananthan , International Advanced Research Centre for Powder Metallurgy and New Materials

-Wearable energy storage devices – Mohitkumar Singh, IIT Mumbai

-Electrically Switchable Fluorescent Polysoft Device, contributed by P.Lakshmi Madhuri , Centre for Nano and Soft Matter Sciences, Contact +91-80-23084200,

Mobile +91-9483070759, email:
lakshmimadhuri17@gmail.com

Grover enterprises, P.L.Tandon &co and Galaxy Lab Instruments are also in this field to supply quality products to create BSL3 and 4.

Grover &Co, 27, Lehna Singh Market, Malka Ganj, Delhi 110007, Phone 9811083291, 8048110679.

Their Fume hood, Laminar Flow, Lab Accessories, Micro Side Cabinet, TLC kit etc are imported from USA, Japan, Korea and U.K. The customers are pleased with the quality and good service rendered by the staff trained by the overseas company.

This reputed company is flourishing for the past 53 years.

Galaxy Lab owned by Rajender Singh Rana deal with lab equipments, BSL3 and 4 with international quality. Their office address 2157T-1 ,Ist floor, Arjun Nagar, Op Shaini Hotel, West Patel Nagar, New Delhi 110008.They have a business turnover of Rs 2 crores annually and they have manpower of 20 only. But they are trained well to satisfy the customers with their sound knowledge in TQM. For the past 15 years they are importing quality goods for BSL 3&4 from USA, UK, and Japan etc. Mobile no: 08377807637.

Indian Health Care and Nanotechnology

Indian hospitals have thousands of in-patients and many more thousands of out-patients.

The capabilities of these hospitals are low and hence they refer to sophisticated hospitals like Apollo, Fortis, and IIMS etc as reference cases. In these hospitals the specialty is that they apply Nanotechnology for treating cardiac patients, bone surgery, wounds, cancer, burns and skin ailments.

Indian government encourages young nano scientists from various institutes to concentrate nano applications for rendering treatment to cancer, bone diseases, skin burns and vision defects etc faced by army men.

Some of the research works and findings are as follows:

Synthesis and characterization of Multi-functional Magnetic Iron oxide Nano particles for Hyperthermia Cancer treatment

As per the words of the scientist Pradipta Rauta "From various Nano Theranostics studies ,Iron oxide Nano particles (IO NPs) have the potential to serve as both nano carrier and imaging agent because of their good bio-compatibility, spatial imaging capability and generate localized heat, when exposed to an alternating magnetic field ,resulting in combined Chemotherapy and Hyperthermia .

The current study aims at developing novel magnet IO NPs and functionalization with various biocompatible surfactants e.g., Oleic acid (OA).

Method adopted

The magnetite (Fe_3O_4) nano particles were synthesized by cost effective co-precipitation method. The physiochemical properties were well characterized using DLS analysis (size and zeta potential), X-ray diffraction (XRD), Fourier Transform Infrared Spectroscopy (FT-IR), Scanning Electron Microscope-Energy Dispersive X-Ray Spectrometry (SEM-EDS) and Vibrating Sample Magnetometery (VSM).

Test results and nano particle size

The results showed the particle size was very much dependent on the Fe3+/Fe2+ ratio pH, initial temperature etc.

The average size and zeta potential of the formulation IO NPs-OA were found to be 35.21±4.23nm and -32.7±5.92 mV respectively. That confirmed increased stability of the nano particles.

The XRD pattern showed the signature peaks for magnetite (Fe_3O_4), suggesting its presence in the synthesized OA-MNPs. The OA functionalization on IO NPs surface confirmed from FTIR study through the analysis of bonding pattern of the Carboxylic acids on the surface of the Nano particles.

Application on Cancer therapy

The uniform size of formulated Nano particles was revealed through SEM analysis. Magnetic characteristic of Fe_3O_4 Nano particles was indicated super paramagnetic properties making them more suitable for Hyperthermia application.

However, more physic-chemical characterization, in-vitro studies like Cellular uptake study, Magnetic Hyperthermia studies etc. are being focused to evaluate the functionality of the system in vitro and in vivo for Hyperthermia applications in cancer therapy."

Scientist - Pradipta Ranjan Rauta

Address: Mazumdar Shaw Centre for Translation Research Mazumdar Shaw Medical Foundation, A-block, 8th Floor# 258/A, NH Health City Bangalore-560099, Karnataka, India

Phone: +91-80-71222274

Mobile: +91-88-95178107

Email: pradipta.ranjan@ms-mf.org

Organization: Mazumdar Shaw Centre for Translation Research

Co- Scientists: Bhaskar Vishwanathan, Komal Prasad C and Aditya Chaubey

Resveratrol Loaded Polymeric Nano particles in Cancer Therapy

Resveratrol (3,5,4'Trihydroxystilbene) (RSV),is known to have anti-cancer, anti inflammatory and anti-oxidant activity , but its application is compromised by its physiochemical properties such as low stability, increased oxidation to heat and light exposure, poor aqueous solubility, short biological half life, rapid metabolism and elimination leading to its low bioavailability.

Polymeric Nano System

Doppalapudi Sindhu, a Nano scientist from Telangana says, "The main objective of the study was to develop a polymeric nano system for the delivery of RSV, capable of maintaining the stability of the highly photosensitive drug and improve the activity of the drug.

Method

For this purpose, RSV loaded poly (lactic-co-glycolic acid) nano particles (RSV-PLGA-NPs) were prepared by emulsion solvent diffusion method. The influence of different processing and formulation parameters on the entrapment efficiency of RSV in the nano particles was evaluated .The results demonstrated that the entrapment efficiency of RSV in nano particles was mostly affected by the method of preparation of nano particles, type of stabilizer used, drug and polymer concentration.

Nano particles

Obtained data indicated that the optimized methodology of preparation allowed the formation of positively charged homogeneous spherical nano particles with nearly 100nm particle size and improved entrapment efficiency.

The release behaviour of RSV from the developed nano particles exhibited a controlled release of about 50% drug release in 3 days and about 80% release till 11th day .The invitro anti-tumoral activity of RSV-PLGA-NPs was assessed using a human breast cancer cell line (MCF-7 cell line) and the results indicate that formulation enhanced the cytotoxic effect of the drug".

Scientist: Doppalapudi Sindhu

Address: Research Scholar, Department of Pharmaceutics, NIPER Hyderabad,

Hyderabad-500 037, Telangana, India

Phone+9140-23073741

Mobile+91-8985123961E
mail:dsdoppalapudisindhu@gmail.com

Organization: NIPER Hyderabad

Other scientist involved: Sailaja Duvvuri, Upendra Bulbake, Rohan Ghadi and Wahid Khan

Combinational Nano-therapeutics to combat multi-drug resistance in Cancer

A nano scientist Dr. Anjali Jain from Hyderabad expresses, "Multi-drug resistance (MDR) is one of the most significant obstacles in cancer chemotherapy and over expression of P-glycoprotein (P-gp) is one of the mechanisms for MDR, which reduces the intratumoral drug concentrations.

Synthetic P-gp modulators /inhibitors were used to enhance the cellular bio-availability of many anti-cancer drugs by blocking the P-gp efflux pump.

Nano particle System

However their utility is limited due to associated toxicity issues. Co-administration of natural origin P- gp inhibitors has been reported to substantially increase the oral bio-availability of many anti-cancer drugs. Further, incorporation of this combination into nano particulate system provides an added advantage of controlled drug distribution and subsides off target side effects of drug .in this study, PLGA nano particles co-loaded with anti-cancer drug and a natural origin P-gp inhibitor were prepared.

PLGA nano particles possess moderate MDR reversal activity on their own, which may provide additional benefit along with drug combination. Drug loaded nano particles with particle size below 150nm were obtained which showed sustained in vitro drug release.

Method used and results

An everted gut sac method was used to study the effect of P-gp efflux on drug transport and an increased uptake of P-gp substrate anti-cancer drug was observed in the presence of P-gp inhibitor.

Vitro cell line study indicated higher efficacy of nano particles compared to free drug solution. These results suggest that the use of combination of P-gp substrate anti-cancer drug and natural origin P-gp Inhibitor in Nano-particulate System would be a promising approach in the treatment of breast cancer".

Scientist: Anjali Jain

Address: National Institute of Pharmaceuticals Education and Research, Hyderabad-500037, Telangana, India.

Phone: +91-40-23073741

Mobile: +91-7207191132

Email:anjalijaindops@gmail.com

Organization: National Institute of Pharmaceuticals Education and Research, Hyderabad

Co –Scientists: Vaddi Sahithi, Eameema Muntimadugu and Wahid Khan

Conjugation of barium ferrite nano particle with-cyclodextrin derivatives for targeted cancer drug delivery application

Method adopted

Misha Mudalali, a Nano scientist from Tamilnadu states "Barium ferrite nano particles are synthesized using barium nitrate and ferrous nitrate using appropriate solvent by self combustion method in the presence of glycine as fuel. The nano particles thus obtained is substituted with the corresponding nitrates of Zn, Co, Cu, Mn, Ni. β -

Cyclodextrin was tosylated using p-toluene sulfonyl anhydride.

The tosyl- β-cyclodextrin was then converted to aminoethylamino – β-cyclodextrin by reaction with 1, 2 – diaminoethane.

PEG was tosylated tetraethyl amine and ethyl acetate with rapid agitation and synthesized PEG amine from PEG-bis sulfonate via nucleophillic displacement.

6-0-Monotosyl-6-deoxy- β-cyclodextrin (β-CDOTs) has been synthesized using β-cyclodextrin and p-toluene sulfonic anhydride by aid of a reticulation reaction.

Nano particles

Conjugation of PEG with tosyl β-cyclodextrin and then the substituted barium ferrite nano particles of different composition have been characterized by XRD, EDAX, and UV Spectroscopy. Their size distribution and morphology were studied using particle size analyzer and scanning electron microscope respectively and their bio-medical application as cancer drug carriers have been investigated.

The particles are found to be of sub-micron size (490nm), which can hold nano drug formulation for delivery at taret site. The different derivatives and their crystal structures were deduced from the XRD pattern.

Scientist: Misha Muthalali

Address: Department of Nano Sciences and Nano Technology,

Karunya University, Coimbatore-641114, Tamilnadu, India

Phone: Mobile: +91-8903773990

E mail:misham.misha@gmail.com

Organization: Karunya University

Co Scientists: I. V. Muthu Vijayan Enoch, Reni E.George and Rithin P.Krishnan

Graphene oxides develop new blood vessels by Angiogenesis

Nano process

Nano scientist Sudhip Mukherji says, "Angiogenesis is a process that helps to make new blood vessels from pre-existing vasculature.

Angiogenesis has an important role in various psychological processes including embryonic growth and development, skeletal development, wound healing etc.

Naturally, healthy body maintains the equilibrium of angiogenesis.

Nano materials

Many cardiovascular diseases such as atherosclerosis, ischemic heart and limb diseases and others such as rheumatoid arthritis and diabetic retinopathy are the consequences of the impaired angiogenesis process.

Conventional treatment strategies include expensive pro-angiogenic molecules (VEGF, bFGF, PDVG etc), which has several limitations. Therefore, design, development and identification of low cost pro-angiogenic molecules are urgently required for the alternative treatment of cardiovascular related diseases for Indian army men. Recently, our group has developed various nano materials, which show excellent pro-angiogenic properties.

Results and treatment

In this context, the angiogenic properties of Graphene oxide (GO) and reduced grapheme oxides (rGO) have been demonstrated, observed by several in vitro and in vivo angiogenesis assays. The intracellular formation of reactive oxygen species (ROS) and reactive nitrogen species as well as activation of phospho-eNOS and phospho - Akt might be the plausible mechanisms for GO and rGO induced angiogenesis. Interestingly at higher doses (>100μg/mL) both GO and rGO showed excessive ROS production, resulting anti-angiogenesis.

These results suggest the ROS dependent switchover between angiogenesis and anti – angiogenesis by GO and r GO, which can be utilized for development of alternative therapeutic approach for the treatment of cardiovascular related diseases and cancer.

Scientist: Sudhip Mukherji

Address: Bio-ChemistryLab-2, Main Building, Indian Institute of Chemical Technology

Habsiguda, Hyderabad – 500 007, Andhra Pradesh, India

Phone: +91-40-27191459

Mobile: +91-7306343460

E mail:sudip.mukherjee1988@gmail.com

Organization: CSIR-IICT

Co Scientists: Pavithra Sriram, Ayan Kumar Barui, Susheel Kumar Nethi and

Chittaranjan Patra.

Chitosan template bimetallic silver Nano particle - Gold Nano cluster Theranostic nano composites for anti-cancer application

Method

As per the Nano Scientist Deepanjali Dutta from Assam "Biopolymer chitosan embedded bimetallic silver Nano particle-gold Nano cluster theranostic module has been developed for cellular application by rapid and simple galvanic exchange method.

Nano scale

Silver nano particles induced ROS mediated cell death of HeLa cells and gold Nano cluster exhibited luminescence with good photo-stability and quantum yield ,which are suitable for bio-imaging . Hence, the composite nano particles offer combinational properties of the two metals present in two different nano scale range of sizes enabling killing and imaging of HeLa Cancer cells simultaneously.

Nano particles

The uptake of the composite nano particles was probed via fluorescence of gold Nano cluster by flow cytometry without the use of conventional organic dyes. The possible molecular pathway of uptake and cell death was illustrated by time dependent TEM imaging of cancer cells and the mechanism of cell death was ascertained by cell cycle analysis deciphering apoptotic mediated cell death for this combined module".

Scientist: Deepanjali Dutta

Address: Nano-biotech Lab, Centre for Nanotechnology, J-Block, Academic Complex, IITGuwahati-781039, Assam, India

Phone: 913612583067

Mobile: 918402093437

Email:deepanjalee@iitg.ernet.in

Organization: Indian Institute of Technology Guwahati

Co-author: Amaresh Kumar Sahoo, Arun Chattopadhyay and Siddhartha Sankar Ghosh

Development of MR functional 3 d scaffold for image guided evaluation of bone regeneration using MRI

Nano technology expert and scientist Dr. Sajesh from Kerala says "In order to implement magnetic resonance imaging (MRI) based in vivo monitoring of bone tissue regeneration; we have developed an osteoconducting MR contrast enabled 3-D scaffold by incorporating MR functional nano Hydroxyapatite (MF-nHAp) into a polymer matrix.

Test results

Paramagnetic property of MF-nHAp could provide MR contrast for the developed scaffold and significantly enhanced contrast between the scaffold and host issue. Incorporation of MF-nHAp into the matrix is optimized considering the bio-compatibility and MR contrast requirement.

Results from the ALP measurement, mineral deposition and expression of bone specific proteins proved the osteogenic potential of the scaffold.

Ex-vivo tissue regeneration of hMSC seeded contrast scaffold, when cultured for 21 days showed distinct changes in the MR contrast with regard to matrix deposition and morphological change. In vivo studies using rat cranial defect also showed significant contrast change corresponding to cell infiltration and tissue regeneration.

Our results suggest that the unique osteoconducting MR contrast enhanced scaffold can be utilized for image guided evaluation of various stages of bone regeneration non-invasively. Our results suggest that the unique osteoconducting MR contrast enhanced scaffold can be utilized for image guided evaluation of various stages of bone regeneration non- invasively".

Scientists: Sajesh K.M

Address: Amrita Centre for Nano Science and Molecular Medicine, AIMS.P.O, Kochi –683041, Kerala, India

Phone: 9148420081234

Mobile: 919745896648

Email: sajeshkm@aims.amrita.edu

Organization: Amrita Centre for Nano Science and Molecular Medicine

Co-author: A.Anusha, G.S Gowd, S.V Nair and Manzoor

CNTFET for Cervical Cancer Detection

Nano scientist B.L.Radha from Karnataka says "Cancer of the uterine cervix is the most common cancer affecting Indian women with an estimated 142000 new cases coming to light every year and 77000 women dying of the disease.

Cervical cancer is the second most common type of cancer in women worldwide and is responsible for the deaths of over 250000 women each year.

In this work the CNTFET device is fabricated and operated under n-channel depletion characteristics .The aldehyde groups from the glutaradehyde linked to the amino groups of

APTES is attached on the biosensor surface to immobilize anti-CSA molecules .The output signal of the developed CNTFET biosensor can be used to detect CSA concentration in the range from 5 to 5000 pg/ml".

Scientist: B.L Radha

Address: Visvesvaraya Technological University, Chikkaballapur-562101, Karnataka, India.

Phone 918085539689334

Mobile: +919482535094

Email: rashablrbl@gmail.com

Organization: Visvesvaraya Technological University

Co-Scientists: Dr.Cyril Prasanna Raj P., Dr.Padmalathavenkatram

Dr Dinesh Rangappa and Ashwath Narayana B.S

Direct electron transfer based aptasensor for the ultra sensitive detection of Cardiac Biomarker Myoglobin

Nano technology expert Munish Shorie from Mohali opines "Detection of Cardiac Biomarkers is an influential part in the triage of patients with dysapnea.

Myoglobin is amongst the first cardiac biomarkers released and can be used for the early detection of cardiac ischemia.

Method and System adopted

Here we present a label free method to detect Myoglobin exploiting its direct electron transfer on an rGO/CNT nano-structured electrode decorated with an aptamer specific to Myoglobin .The developed system exhibits synergistic effect combining the electrical properties with the chemical functionality to aid the development of dual rGO/CNT bio-interface.

The Mb specific aptamer generated by modified SELEX showed superior affinity (k_D- 65pM) thus imparting the developed system with its selectivity and high sensitivity .The aptamer functionalized rGO/CNT electrodes demonstrated a dynamic signal response between 1ng/ml to 4µg/ml with a detection limit of ~0.34 ng/ml .

The newly developed DET based assay presents a promising candidate in point-of-care diagnosis for screening of Mb in patient samples".

Scientist: Munish Shourie

Address: Institute of Nano Science and Technology, Habitat Centre, Sector -64, Phase 10, Mohali -160062, Punjab, India.

Phone: +91-1722210075

Mobile: +91-9056198030

Email:munish_shorie@yahoo.com

Organization: Institute of Nano Science and Technology

Co Scientists: Ashok.K.Ganguli and Priyanka Sabherwal

Development of MEMS Micro-Cantilever Sensors for Detection of DNA Single Nucleotide Polymorphism in Cardiac Point of Care Diagnostics COA Therapy

Aviryu Kumar Basu, a Nano expert from Uttar Pradesh states "Optimal anticoagulation treatment is required for patients with increased risk of Thromboemboli formation leading to Thrombembolism.

Such Thrombembolism can occur after heart value replacement in patients suffering from a trial fibrillation, patients suffering from deep vein thrombosis. In all these cases it is important to administer the COA Therapy in post surgical phases.

The principal mechanism of COA therapy is the interference with the Vitamin K cycle in human being. In the Vitamin K cycle the VKORC1 enzyme and γ-Glutamyl Carboxylase (GGCX) simultaneously converts Glutamic acid to γ-Carboxy Glutamate, which results in biologically active calcium dependant clotting factors II,VII,IX and X.

An international normalized ratio INR has been adopted by the World Health Organization (WHO) to measure the Prothrombin level of patients, who are at risk to Thrombosis

Short nucleotide polymorphism (SNP) in gene coding of the COA metabolism enzyme (VKORC1 and GGC), it has been found that the Polymorphisms in CYP2C9 gene has been associated with an impaired ability to metabolic COA. Hence prolonging drug half life and leading to reduce dosage requirement .Therefore the screening for the gene mutations using SU-8 cantilever before initiating COA treatment may reduce the risk of bleeding and haemorrhage.

Cantilevers of various designs are developed through lithography in a simple 2-step process thickness in the range of 2-4 um with a simple peel off process from the substrate".

Scientist: Aviru Kumar Basu

Address E1-214 HALL-4 IITK Kanpur-208016, Uttar Pradesh, India

Phone: +915122596056

Mobile: +918765674643

Email: aviru@iitk.ac.in

Organization: IIT Kanpur

Co-Scientists: Prabhat Dwivedi and Shantanau Bhattacharya

Corneal Inflammation Sensing

Gelatin Nano particles with Host Regulated Anti-microbial and Anti-Inflammatory responses

Saad Mohammed Asan from Telangana, Andhra Pradesh says "Corneal infection/ inflammation (Keratitis) is considered to be a major cause of avoidable visual impairment in the world. an important issue in the management of Keratitis is the requirement of maintaining therapeutic concentrations of two different drugs (anti- inflammatory and anti- microbial), in an appropriate balance, over prolonged periods.

Moreover, the various anatomical and physiological barriers that normally protect the eye create obstacles during therapeutic interventions, making the treatment even more challenging.

System

The purpose of the present study was to develop a nano particle based system to overcome the shortcomings in the current procedures for Keratitis treatment.

The nanostructure system described here exploits two hallmark feature of the Keratitis patho-physiology (over-expressed TLRs and increased protease activity) to modulate drug activity.

Nano particles

The Nano particle consists of a Ketoconazole (ket) –loaded gelatin core and surface conjugated anti- TLR4 antibodies. While the anti-TLR4 anti-bodies assist in sensing infection/ inflammation, corneal adhesion and suppressing inflammation, the gelatin core degrades in response to the protease activity for an on- demand Ketoconazole release. The nano particles synthesized were successfully tested on human corneal epithelial (HCE) cell culture models and rat models of fungal Keratitis.

The present study addresses basic issues in corneal drug delivery and demonstrates a simple yet smart drug- delivery system with host controlled anti-microbial and anti-inflammatory effects for an effective management of fungal Keratitis.

Poster presenter: Saad Mohammad Ahsan

Address: W-209, West Wing 1st Floor, CCMB Uppal Road Hyderabad-500007, Telangana, India

Phone: 914027192556

Mobile: +919959643017

Email: saadmahsan@ccmb.res.in

Organization: Centre for Cellular and Molecular Biology

Co Scientists: Ch.Mohan Rao

These innovative ideas given by scientists can pave a new way for health care management for the welfare of our defence sector and the public at large.

Purpose, System and Materials used for Production of Nano Composites!

Thermoplastic resins are used for toys, domestic applications, automobiles etc. But thermosetting resins are used for marine, aerospace and defense equipments in India.

In India, much concern is about the usage of nano composites materials for the following purpose and sectors:

1. Anti corrosion coating

2. Superior strength fibres and films

3. Lubricants and paints free from scratches

4. Fire retardant materials

5. UV protection sets

6. Auto industry- door handles, mirror housings engine covers

7. Domestic appliances like Vacuum cleaners, Mobile phones, Electronic equipments, Head phones etc.

It is significant to note that due to incorporation of small quantities of nanoclay materials, the gaseous barrier capabilities have been enhanced. The experts have revealed that oxygen transmission rates for polyamide organ clay composites are lower than half of the unmodified polymer.

The total barrier performance is found to be better due to the amount of clay incorporated in the polymer and the aspect ratio of the filler.

In fact the aspect ratio plays a vital role for betterment of gaseous barrier.

Much research is being done in Indian laboratories for active passive barrier system for Polymide-6 materials.

To achieve passive barrier phenomenon, the nano clay particles are incorporated by nano clay through melt processing methods. The active contribution is achieved by an oxygen scavenging ingredient.

The nano clay particles offer greater tortuosity and bring down the transmission of oxygen through the composite. It also drives molecule to the active scavenging species and achieves zero oxygen transmission.

This great phenomenon has attracted food packing industry to use it.

Fruit juices, confectionary, cheese, dairy products, soft drinks and paper board packets utilize nano composites materials for good period of shell life.

Data has revealed that incorporation of nano clay has the capacity to lower the solvent transmission through polymers like polyamides. There is considerable quantity of reductions in fuel transmission via Polyamide - 6/66 Polymers by incorporation of nano clay fillers.

So, the Indian industrialists have shown interest to use these nano composites materials in fuel line components of auto mobiles. This has supported the cost of production of the concerned cars, trucks etc.

Instead of using the conventionally filled polymers, the addition of nano clay supports the transparency qualities, which is needed for the films.

Modifications in the crystallization nature is caused by nano clay particles, say the scientists.

Nano modified polymers have the capacity to increase the toughness and hardness of polymeric transparency material, when used for coating. It is also important to note that they do it without disturbing the light transmission behaviour. The nature of resisting high velocity impact along with increased abrasion resistance is also seen.

Usually polymeric materials are affected by water laden atmosphere. So, the quality to reduce the water absorption is a must, which can be attained by inclusion of nano clay.

The same effect can be obtained from polyamide based nano composites. Due to the enhancement of aspect ratio, the absorption of water is reduced. This is the special advantage of nano particles over the addition of micro particles.

The reduction in flammability nature in polymers can be achieved by incorporation of nano clay. The usage of conventional micro particle filler with flame retarding agents may reduce the flammability phenomenon. But it will disturb other vital characteristics. But it is not so in the case of incorporation of nano clay.

The production cost is high and when nano composites are used a team of experts is needed to execute it.

Innovation is lagging in India and knowledge about the impact of the performance, formulation relationship (which is complex in nature), ways to attain and measure nano fillers dispersion and exfoliation in the polymer matrix are still worrying the scientists in India. They have to update in their research works in future.

"In later stage, new types of polymer system have to be evolved and the scientists have to equip themselves with new strategies to tackle the compatibility aspects. They are still working on PVC based systems and PET nano composites. Further reinforcement of clay nano composites by glass fibre is to be studied in depth by Indians. They also now focus their attention on production of electrically conducting clay nano composites" remark the experts.

"For high valued abrasive materials or wear resistant coating materials we can consider the metal nano composites as the best choice. Metal nano composite powder is used for aerospace, medical and for machine parts with high performance. Its capability to become bulky at faster rate is an added advantage. High quality sintering performance of it compared to conventional metal composite materials is a special feature for usage in industrial or defence applications. Whenever less weight is needed as in defence equipments, the lower density polymeric materials are the best. By including thermally and electrically conducting fillers it can also develop nano composites, with electrical conducting characteristics" remark the scientists.

Automobile industry uses the thermoplastic nano composites for under hood applications and interior parts. For the plastic wrap and carbonated beverage containers, this is found to be suitable.

Thermoplastic and thermosetting have the differences in process cycle, structural components, recyclable reprocessing nature, durational of the process cycle etc.

Thermoplastic resins are used for toys, domestic applications, automobiles etc. But thermosetting resins are used for marine, aerospace and defence equipments.

To produce thermostat nano composites integration of nano particles with polymers is done. These are complex hybrid materials with quite different properties and characteristics for various industrial production activities.

The Indian researchers do not show much care for thermostat nano composites compared to thermoplastic nano composites. At present some have taken steps to produce it with updating of knowledge on characteristics of the inter phase-region and the calculation of technology-structure property relationships. Many nano particles are used to prepare nano composites with thermosetting polymers along with diamond, alumina, graphite and smectite clay.

These thermostat nano composite gives better thermal stability, chemical resistance and flame retarding capabilities. Hence marine industries, construction field etc engage it to fetch better customer satisfaction. They have the advantage of shaping with injection moulding and extrusion processes.

In our country these materials are used for packing and fibre industries also.

****

Scientists and Institutes participated in Bangalore India Nano 2016

It is a sorry state of affair that Nano Technology is considered as a gimmick by the illiterate Indians and hence the industrialists are not prepared to introduce it in their production agenda. They feel that people will not agree to what they say about Nano.

The Bangalore India Nano Conference held in Hotel Lalit Ashok, Bangalore from 3rd March to 5th March 2016 was organized by Jawaharlal Nehru Advanced Scientific Research and Karnataka Science & Technology Promotion Society and Department of Science & Technology, Govt. of Karnataka.

This great event was excellently supported by Vision Group on Science & Technology, Nano Mission, Department of Science & Technology, KBITS, R&D Wing of Indian Oil Corporation and Visvesvaraya Technological University in Belgravia.

More than 600 delegates from all over the world like Canada, Illinois, USA, South Korea, U.K and Paris graced the conference conducted on 10 acre land of that 5 Star Hotel.

Research students from IISC, IIT (Kanpur, Mumbai, Chennai and Delhi) University College, Palayam (Trivandrum) Karunya University (Coimbatore) etc. participated with presentation of papers and ideas on Nano technology.

Totally 93 new inventions were brought to lime light and were honoured by the organizers. Professor C.N.R. Rao, Chairman of Karnataka's Vision Group on Science & Technology and Hon President of JNCASR was the Chief Mentor of this grand conference.

He frankly said that "In India much work is done on research works but nothing is implemented. No one comes forward to commercialize it. The research paper on Super Capacitor was first submitted by me several years back and pitiably nobody was willing to introduce it in the market. While Indians neglected it totally, now an USA entrepreneur had at last taken it."

It is a sorry state of affair that nano is considered as a gimmick by the illiterate Indians and hence the industrialists are not prepared to introduce it in their production agenda. They feel that people will not agree to what they say about nano. Scholarly Scientists alone satisfy themselves with the benefits of nano and not the public. Hence these companies are interested in export and not for local market business.

Topic covered by Young Hee Lee, Director for Centre for Integrated Nano Structure Physics in South Korea was 'Unique Features of 2D Layered Materials'.

Prof. Dr. Clement Sanchez from Chemistry of Hybrid Materials College de France –UPMC, Paris spoke about 'Nano Structured Hybrid O-I and Inorganic Materials'.

The Polarized Electrode Third generation of Capacitive De-ionization was the topic covered by Mr. Marc Andelman from Greater Boston.

'Low Cost Arsenic Bio-Sensors for Point of Use Detection in Drinking water' was extensively enlightened by Prof. Tony Cass, Department of Chemistry, and Imperial College U.K.

Prof Sung Jae Kim vividly exposed the ways to overcome water shortage problems through his analysis of 'Capillary Ion Concentration Polarization as Spontaneous Desalting Mechanism'. This is a power free approach and suitable for developing countries like India.

Regeneration of skin, nerve and blood vessels was the topic covered by Dr. S. Swaminathan, Centre for Nanotechnology and Advanced Bio Materials, SASTRA University, Tamilnadu, which ideally suitable for treatment to wounded soldiers. The Indian army men are badly injured after hectic fight with terrorists and his solution to regenerate infected heart, skin and development of theranostic materials for pancreatic cancer seen in many ex-servicemen.

Another vital need for defence men was to boost the battery life of mobile phones. Prof. V.Ramgopal Rao, Chair Professor of Nano technology IIT, Mumbai, gave a solution for it with small switches, which can bring down power leakage in integrated circuits and thus boost the battery life of mobile phones handled by BSF in remote areas.

Even to detect the explosives placed by terrorists in busy markets, temple festivals, 5 star Hotels and car parking areas, he had mooted the idea of using Nano Electro Mechanical Switches developed in IIT Mumbai. To sniff like a dog and detect the culprits is his aim to save public from anti-national elements.

The Indian defence men should shoulder with him and see that all over India his invention is implemented to save the public from calamity. For this invention and approach to save this country from enemies he has been awarded Prof. C.N.R. Rao Bangalore India Nano Award 2016.

His device can enter the area where dogs cannot enter and detect the hiding culprits in any building or crowd. The disasters happened in Mumbai has provoked him to put an end to further bomb attacks from terrorists.

He hails from Mumbai and feels the pinch about the fear which still exists in the minds of Mumbai residents. Prof. S. Karthikeyan, Department of Materials Engineering, IISC Bangalore, focuses his attention on defence and aerospace through his study on engineering metals at extreme high temperature, alloys and inter-metallics.

He obtained his Doctorate from Materials Science and Engineering, Ohio University.

Dr. Geetha Balkrishna, Director for Centre for Nano and Material Science, Jain University Jain Global Campus, Bangalore, member of the Board of Research in Nuclear Science, Nano Mission and Association of Hazardous Materials etc. was in the conference.

Various Institutes and Scientists in Nano India 2016 Conference

1) National Research Development Corporation New Delhi headed by Dr. H. Purushothama, This is a Central Public Sector undertaking, guided by the Ministry of Science & Technology Govt of India, to support Nano Technology.

2) Nano Science and Technology Consortium (NSTC): This is a division of Consortium Learning Network P Ltd., Noida.

Dr. Puneet Malhotra is the Founder Director of this organization, which aims to enlighten Indian youths about nano technology and its applications for various fields including defence.

Dr.Puneet has the credit of encouraging entrepreneurs to take up business based nano technological researches. This will help indigenization of defence equipments and will also arrange tie-ups with foreign companies involved in these nano composites.

3) Centre for Nano Science & Engineering Bangalore involved in researches in self cleaning walls, clothes, non-icing surfaces for aircrafts anti-microbial surfaces for health care.

Prof. Prosenjit Sen, Assistant Professor in this reputed institute has done research on drug reduction in ships though creation of surfaces, which can resist liquid and particulate contaminants.

4) National Centre for Excellence in Technologies for Internal Security, Centre for Excellence in Nano Electronics, National Centre for Excellence in Nano Electronics, National Centre for Photo Voltaic Research and Education etc. are engaged in large scale production

of Nano materials and their applications for healthcare, environment and defence. Professor Anil Kumar serves in these reputed institutions.

5) Centre for Nano Science and Engineering, Bangalore is headed by Prof. Rudra Pratap, who has served in Sibley School of Mechanical and Aero Space Engineering, Cornell University.

Research works are done on Energy Dissipation in Dynamic MEMS Devices

6) R&D wing in Indian Oil Corporation: This government owned organization is involved in usage of Graphene, which has great thermal conductivity at room temperature. Dr. S.S.V. Ramakumar is the Executive Director, who has been awarded by Department of Atomic Energy for his works on Marine Oil Development.

Prof. A.K. Sood, Department of Physics, IISc Bangalore is an Honorary Professor in JNCASR. Bangalore. He is the Vice President of Indian National Science Academy and Secretary General of the World Academy of Sciences. He does research in Nano Systems and Soft Matter.

He has been awarded by DAE, JNCASR and FICCI. Prof T. Pradeep, Department of Chemistry IIT, Chennai, focuses on clean water with nano technology. Novel methods to eliminate heavy metals, radio nuclides, pesticides and pharmaceuticals are his contributions in this field.

Prof. Dr. Shantikumar Nair, Professor and Director of Amrita Centre for Nano Science and Molecular Medicine at Kochi, has received many awards for his contributions in Nano Science, Nano Composites and Material Science research projects.

Dr. Ashok Mulchandani is a reputed scientist on Homeland, Security and Environmental Care apart from Devices and Sensors for detecting Biological elements.

He has developed CNTs, Graphene and 3 dimensional Graphene. His papers have been published by 275 journals on subjects like Nano Field Effect Transistor Based Bio-Sensors for Healthcare.

He is the Professor of Chemical and Environmental Engineering, Bourns College of Engineering, University of California USA.

Prof. CNR Rao, recipient of Bharat Ratna award from Government of India has published 50 books and presented 1650 research papers on Solid State and Material Chemistry. He has secured doctorate from 69 Universities. He is the member of popular Science Academies in Russia, U.K. USA, Canada, China and Europe. He has received hundreds of medals and awards all over the world.

He is the Chief Mentor of all Nano India Conferences conducted for the past 8 years.

Dr. G. Sundarajan has specialized in the subjects like Electro Spark Coating, Micro Arc Oxidation Coating and EB-PUD coating. More than 36 private organizations have adopted his research works and technologies. As a reputed scholar in material science he has presented papers on erosion, abrasion of metallic materials and composites. Laser materials processing is also another arena, where he has been awarded by National Science Academy American Ceramic Society etc.

Research Industry Collaboration in this Expo enabled young scientists to present their innovations to attract Indian and foreign investors in Nano technology. Focused Tutorials arranged by this Expo also enlightened the modern techniques and their significance in Nano field. Interactions with Indian scientists with delegates and industrialists showed a way for business boost.

This Nano India 2016 conference was totally exciting with scintillating speeches and colourful ideas of distinguished scientists.

Hats off to the organizers!

****

Exhibitors in India Nano 2016

The grand Nano India 2016 Conference in Bangalore was made interesting and informative with display of Nano products from 27 exhibitors and presentation of 93 posters by young scientists, based on innovative ideas on Nano technology. Thousands of visitors viewed it in Maghadh of this luxurious hotel in Bangalore and admired it.

Optical Spectroscopy, Custom Gratings and VUV Beam Lines, Forensics, X-ray Fluorescence, Detectors, Atomic Emission, Raman Analysis, Particle Characterization and Surface Plasma Resonance Imaging were displayed by Horiba India P Ltd. They have their head quarters in Kamadhenu, 17/1-32, Bennarghatta Road, Audugodi, Bangalore 560030. Phone: 080-22210071. In Pune and Delhi they render sales and services with a team of experts.

All their products are applicable for Thin Film Metrology, Bio-Chemical, Carbon Materials, Cosmetics and Pharmaceutical field.

The LA960 Particle Size Analyzer they deal with has 87 detectors, with measurement range of 0.01μm to 3000μm and in-line Ultrasonic Probe.

They also deal with Laser Diffraction Particle Size Analyzer, Dynamic Light Scattering Zeta Potential Nano Particle Size Analyzer, Carbon, Nitrogen and Hydrogen Analyzers.

Contact person: Umesh Krishnamurthy.

Cell No 9686922933.Email:umesh.krishnamurthy@horiba.com.

Their manufacturing unit is in Plot no. 26, Sector 7, I.I.E SIDCUL, Haridwar, 249403. Phone: 1334 239139. They have an excellent marketing team to satisfy their customers in India and all over the world.

Inventys Research Company P Ltd. located in 208-503-504, Keshava BandraKurla Complex, Bandra (E), Mumbai-400051.

Phone: +91-22-32263300, Fax+91-22-3916-7447 Mobile:- +91 9320013300. This reputed company deals with Particle Online Analysis, Particle Sampling, Counting, Standards Characterization, Viscometers, Polymers for cleaning, Fuel Cells, Ultrasonic Liquid Atomization and Micro Fluids.

They have tie-ups with USA, New Zealand, France, Switzerland, Japan, Italy, Germany and U.K.

Contact person is Snehal Kamble. Cell: + 91- 932001 3364.

Email: skamble@particlescience.biz

This company possesses OHSAS1800:12007, ISO14001:2004 and ISO9001:2008 Certification to exhibit their quality.

Particle Shape & Size Analyzer supplied by Ambivalue (Netherlands), Ultra High Speed Compressors supplied by Celeroton (A.G. Switzerland) Rotary Discs, and Spray Nozzle Atomizer supplied by Oh Kawara Kokohki Co (Japan), Particle Counters, Air Velocity Meters, IAQ / Gas Monitors and Dust Monitors supplied by Kanomax Particulate Sampling Systems (pharmaceutical and environmental) supplied by Westech Scientific Instruments (U.K.) are the products they deal with.

Particle size Analyzer, Image Processing & Analysis Software, Plant Science Modules and Gel Analysis are supplied by Expert Vision Labs P Ltd in India. Contact address: H 202, Ranjit Studio Complex, D.P. Road, Dadar West, Mumbai 400028.Contact person: Mr.Bhanu, Phone: +91-022 4066 4242.

Shimadzu Analytical (India) P Ltd had a stall in this show to exhibit their Multi-purpose Tray Operational Panel, Jog Controller, Simple Screen and Table Top Models with safety standards. They have their office in Sethi Chambers, Plot No.2, D.D.A. Local Shopping Centre, M.O.R. Land, and Near Rajender Nagar in New Delhi: 110060. Phone: 011 41537918, Fax 011-28742124, email: infodelhi@shimadzu.in. They have their branches in Chennai and Mumbai.

KAS Technologies displayed their Ocean CVD. This has the special features like no reactor cool down process, loading, unloading etc. This is a unique invention of Centre for Nano Science & Engineering at IISc along with KAS Technologies. This is considered as the first Integrated CVD for Graphene growth. It has in-built purifiers to eliminate gas impurities.

To change the outlet concentration, KAS Technologies has produced Mixed Gas Generator. It avoids the locking up space and cost as it lowers the dependence on gas suppliers and logistics of holding various combinations of mixed gas. Storage of dangerous gas inside the complex is eliminated. Contact person for any business dealings is Ankur Singh, +91-9900058774, email:ankur@uhptech.com. They have their head office at UHP Technologies P Ltd., E-43, Kailash Industrial Complex, Hiranandani Link Road, Vikhroli, Mumbai, 400079. Phone: 022-40156050, Fax: 022-40156054. They have their branches in Bangalore and Middle East. They have tie-up with DRDO, ISRO and Ministry of Defense, New Delhi.

Their KAS CAB with UHP supports to use gas at multiple points at various flow rates and pressure. We can view the pressure with wire meshed view glass. Another feature is the fire retardant enclosure.

Silicon AFM Probes from Budget Sensors located in Innovative Solutions Bulgaria Ltd. 48, Joliot Curie Str, 1113 Sofia, Bulgaria, drew the attention of many delegates. Single Hi-Res AFM Probes, Magnetic AFM Probes, Conductive AFM Probes and Silicon AFM Probes are their products. Single Wall Carbon Nanotubes introduced by OCSiAl has been successful in battery applications. Aluminium coated with single wall carbon Nanotubes reduces the interfacial resistance.

UV Curers (Ultra Violet Exposure Chambers) was displayed by Apex, which produces micro and nano thin film research equipments. Their office is located in B2C,

Ibrahimpur Road, Jadavpur, Kolkata-700032 (West Bengal). Contact: Phone: +91-33-24130364, +91-33-2429-4401, Fax: +91-33-4065-4054. Their equipments are capable for instant, fast cleaning and curing of thin film coatings as well as for experimental liquids. The customers report that they are user friendly with 2.2W UV output and UV wave length of 254nm. They are also involved in researches to create UV Curer with inert gas controlled atmosphere and in-situ stage temperature controller, UV Curer with Inert Gas Controlled Atmosphere, Automatic Programmable UV curer etc.

Contact person for sale: Anirban Talapatra. Cell: +91-8582849326. E-mail: sales@apexicindia.com.

"Apex has the pride of introducing advanced and unique programmable spin coating system with in situ programmable temperature controller. It is user friendly and has compact shape to occupy minimum space in the laboratory", said their sales team.

UHP Technologies has leading clients like SSPL(Delhi), IISC (Bangalore), Jupiter Solar(Himachal Pradesh), MTW Group, i-clean (Bangalore), Engineers India Ltd.(Delhi), Ellipsiz (Singapore), SABIC (Bangalore), Clariant, KFUPM(Saudi Arabia), BEL (Bangalore), Bayer Material

Science, BHEL, Moserbaer, Indo Solar, IIT (Kharagpur, Delhi, Mumbai, Jodhpur), BARC(Mumbai), IIST (Trivandrum) etc.

SPM probes & Test structures from µMasch attracted the attention of the visitors in this 8th Bangalore India Nano exhibition. They have their headquarters at Bulgaria and branches in Germany and USA.

Low Noise Conductive Silicon Probes, High Resolution Conductive Silicon Probes, Non-magnetic Non Contact Silicon Probes hardened DLC Coated Silicon Probes etc were some of their products with international standards.

Mittal Enterprises is located in 215/T-7C, New Patel Nagar, New Delhi-110008, Phone: +91 1125702784, Fax: 25120261, Mob: +91 9810681132.

Email:mittalenterprises@bol.net.in,info@mittalenterprises.com.

They are the leading manufacturers of laboratory & research equipments in Delhi.

Using their Interferometers several papers have been published by leading Nano technology scientists and experts like:- Magnetic and Ultrasonic Investigations on Magnetite Nano fluids, Characterization of Nano Materials,

Ultrasonic Properties of Nano particles Liquid Suspensions, A study of ZnO Nano particles and ZnO-EG Nano Fluid.

Contact person: Ravi Mittal, who obtained MSc Physics from IIT Delhi and MTech from Delhi University. It is an ISO 9001: 2008 CO certified company.

Reinste Nano Ventures P Ltd. has its sales office in 4, First floor, CSC Market, Pocket-E, Mayur Vihar, Phase-2, NewDelhi, 110091. Registered office is in: 40, National Park, LGF, Lajpat Nagar, IV, NewDelhi. Email:info@reinste.com. Mobile. +91 9810662669. Landline: +91 1204781200. They are engaged in supply of Nano materials like Gold nano particles, Silver nano particles, Fluorescent nano crystals, Gold Nano particles Optimization panels etc. Contact person is Puneet Mehrotra, Director, Administrative Office, A-118, 1st Floor, Sector 63, Noida-201301, UP. Phone: +91-120-4781201, Mobile +91 9810367262.

Semiconducting Nano Wires, Magnetic Nano Particles and Support Reagents supplied by them drew the attention of the visitors. They also deal with Kisho Nano Glass Coating, which is UV absorbent, anti-fouling, highly durable, anti-dust and anti-scratch, said their representatives. So, all food packaging, electrical, plastic and paint manufacturers seek their products for better quality and enhanced customer satisfaction.

Nano glass coating is a silicon dioxide based coating, which has many appealing features, like water repellent and heat resistant up to 1000°C.

All these stalls were visited by dignitaries and foreign investors interested in devices meant for biological agents.

Trade Fairs attended by Nano Composite Scientists and Subjects Covered

International Conferences on advanced Nano Materials and Emerging Engineering Technologies held in Sathyabhama University were attended by leading scientists from all over the world.

The Centre for Nano Science and Nano Technology is in existence for the past 10 years in the Sathyabhama University campus, with the aid of IGCAR.

Scientists in it are engaged in various conferences based on Nano materials, Nano Composites, Nanofabrication, Nano Electronics and Nano Technology conducted in this campus. They invite foreign experts, consultants and industrialists interested in implementing it. Serious discussions across the table take place in this University Conference Hall.

Papers are submitted on subjects like Synthesis of Nano Materials, Nano Surface Modification by Ion Implantation, Nano Semiconductors, Physics of Carbon Nano Tubes, Synthesis of Nano Particles, Processing of Polymer Nano Composites, Nano particles Surface Modification, Polymer Nano Composites Characterization Behaviour and Performance, Synthesis of Thin films, Micro Structural and Compositional Analysis, Thin Film Coating to Control Bio Fouling, Bio Nano Materials, Bio Nano Sensor, Nano Electronic Devices, VLSI and Nano Electronics, Anti Microbial Activities of Nano particles, Micro Structural and Compositional Analysis etc.

The University has spent crores of rupees by imparting vital research equipments for the scientists to conduct advanced research projects. They are Optical Microscope, Ultrasonic Bath, Bacteriological Incubator, Dispenses Bottles, Epiflourescense Microscope, Atomic Force Microscope, Field Emission Scanning Electronic Microscope, E beam Evaporator etc.

All equipments are imported from countries like Italy, France, Japan, Ireland, Japan, Denmark and Germany.

International Conferences on Advanced Nano Materials and Emerging Engineering Technologies held in

Sathyabhama University were attended by leading scientists from all over the world.

National Workshop on recent trends in X ray Diffraction Techniques, International Conference on Energy Materials, International Conference on Recent Advances in Physics for Inter Disciplinary Developments, National Seminar on R&D in Nano Sensors, International Conference on Advanced Nano Materials and Emerging Engineering Technologies, National Conference on Trends in Renewable Energy Sources Applications and Technologies, International Conference on Nano Science and International Conference on Emerging Trends in Robotics and Communications are some of the interesting conferences visited by them.

Workshop on Advanced Thin Film Techniques, National Nano Technology Meet on Energy and Environment, International Conference on Emerging Trends in Robotics and Communication Technologies sponsored by DRDO were some of the leading conferences conducted here and many scientists met for discussions recently.

The scientists show keen interest to participate in congresses and projects on following subjects:

Subjects

1) Novel Metal Complex Photo Catalyst System for Carbon dioxide Splitting

2) Synthesis of Carbon Nano Tubes for Gas Sensor application (Sponsored by IGCAR-DAE).

3) Drug delivery in Zebra Fish Model.

4) Development of Carbon Nano Tube based Epoxy Composites (with the aid of ADE-DRDO).

5) Infectious disease in Zebra Fish Model with the aid of DBT- IDB - Medical Biotechnology.

6) Bio-Waste Modified Concrete Structures.

7) Coatings of Super Alloy for Nuclear Wastes (supported by UGC DAE-CSR).

8) Ballistic protection (supported by DRDO).

9) Thermal Barrier Coatings.

10) Investigations of Microbiological aspects of Bio Deterioration / Bio-degradation of Concrete of Sulphur Oxidizing Bacteria/Fungi under the patronage of BRNS.

11) Improving the Anti-microbial properties of Condenser Material by Surface Modification using Nanotechnology with the technical backing by IGCAR-DAE.

For all these International Congresses and Trade Fairs, the economically backward Indian scientists cannot afford to spend from their pockets to manage travelling expenses, accommodation and delegate fees charged heavily by the organizers. So, the leading institutes and MNCs have to share the financial burden to encourage the scientists to present papers and take part in discussions regarding the applications of Nano Composites for industrial production to raise the GDP. Wide publicity had to be given to all institutes through newspapers and notice boards to reach the scientists. They don't have associations or unions to raise their voice to be considered for participation. Hence the individual has to toil, run here and there to know about the conference details, fees and schedule. No bank in India supports their participation in Trade Fairs and Congresses even though the public sector banks give liberal loans for higher education. Non-refundable loans are given for farmers but not for the scientists to attend these useful meets.

These aspects are to be considered to back up innovations and researches in India.

Companies, Institutes and Scientists involved in Nano Composites

Advanced memory devices, lighting gadgets, Bio sensors for medical checkups and other defense research works based on nano composites are carried out in leading Universities like SRM, IIT, Indian Institute of Science Bangalore, IIMS (Delhi), SASTRA (Tanjore) and Sathyabhama University in Tamilnadu.

Nano technology has created a revolutionary scenario in India.

Many clamor to know about it, as it is yet to reach the common man and the public. The government's efforts to enlighten the young and budding entrepreneurs to take up manufacturing efforts are the order of the day. Only MNCs and Research Institutes with sound financial background and sophisticated R&D set up, take a plunge in it. Others hesitate as the common man in the streets will not respond to it and the investment will lead to loss commercially.

Companies

BASF play a vital role in nano composites concerned for defense applications. They focus their attention on nano particles with highly branched Polyisocynates for coatings and camouflage. Their coating will give high quality abrasion resistance and anti-reflection, required refractive indices and prevention from corrosion. In addition to these they render surface with self cleaning in nature.

So, the defense sectors highly recommend their products for the army men's weapons. Indian government seeks the support of the polymer researchers in that internationally reputed BASF.

Again the Nano experts in this reputed company have won laurels by experimenting in nano cubes created out of organ metallic network materials. This will be a vital storage medium for hydrogen. It is achieved by their 3 dimensional lattice structure, which passes many pores and channels.

As automobile experts gain customers through low weight automobile parts fitted into them, Toyota is popular in utilizing the Nano composite technique in their bumper and advertises that they are anti-scratch anti-dent. Another automobile MNC is Chevrolet Impala, which utilize polypropylene side body molding reinforced with Monthorillonite.

This technology used by them reduces weight considerably and increases hardness. Oxonica has the pride to use cerium oxide for betterment of combustion.

These aspects for fuel economy are a must in military agenda.

Kodak is another MNC, which use nano structured polymer to produce organic light emitting diode and color screens for car stereos and mobile phones.

Angstro Medica, a popular MNC engaged in producing nano particulate based synthetic bone, has drawn the attention of all physicians and ortho specialists. They have manipulated calcium and phosphate at the molecular level and developed bone similar to our natural God made bones. They have created vaccines and ailments for genetically based illness by mixing Genomics with Nanotechnology.

Antibacterial dressing covered with nano crystalline silver is created by M/S Smith & Nephew. "This has the capacity to destroy wide variety of bacterial germs within a short period of 20 minutes", say their experts.

Institutes

The flow induced electrical response in CNT has good relevance in biological and biomedical projects, say the nano experts in Indian Institute of Science. This technology has been offered to a foreign firm for gas flow sensors.

In Pune, we have the popular National Chemical Laboratory, where the talented nano scientists have created nano crystalline gold triangles to attend cancer patients by Hyperthermia. The irradiations of the cancer cells are carried out by infrared radiation, explain the medical stalwarts.

Their contribution is vital for preparing insulin delivery for advanced diabetics.

Banaras University is another leading University in North India, which is studded with nano experts, who have created a membrane out of CNT to remove bacterial germs from water. Their efforts are successful and appreciated by the customers, who use that water for drinking purpose.

To eliminate the dissolved pesticides in drinking water, the scientists from Indian Institute of Technology in Chennai, Tamilnadu have formulated nano silver based water filter. Eureka Forbes has introduced this unique water filter to their millions of Indian customers. Especially in villages, their products have won appreciation, as water in most of the bore wells and ponds in the villages is mixed with pesticides.

Hyderabad is another important city, where Bio-technology Park has grown with many stalwarts. In that city we have International Advanced Research Centre Powder Metallurgy and New Materials, which has the capacity to generate many varieties of ceramic, metallic and cement nano powder. They have the required synthesis facilities with the imported technology and support from foreign university experts.

They also have the pride to produce low cost nano silver coated ceramic candle to eliminate bacterial germs from tap water and well water which is used for drinking and cooking. Their scientists have also worked to produce lightning arrestors with the help of ZnO Micro Crystalline Powders.

Polymer nano composites will play a vital role in research institutes in India. Previously the Indian experts were involved on plastic products and chemicals for domestic and industrial usage. But now their full focus is on PNC only.

In India, the industries for aerospace, packaging, solar cells and electronic goods have to change over to PNC concepts for better usage. New varieties of products are the dream of every Indian, since they reduce the cost. As Indians are economically backward any cost cut will attract millions of customers quickly.

The elites prefer the introduction of nano composite materials in their cosmetics, beauty parlor products and sophisticated lightweight automobiles with luxury fittings.

Computer chips out of this technology will play a vital role, as computer has entered each and every village, town and city of Indians. Advanced memory devices, lighting gadgets, bio-sensors for medical checkups and many defense research works based on nano composites are carried out in leading Universities like SRM, IIT, Indian Institute of Science Bangalore, IIMS (Delhi), SASTRA (Tanjore) and Sathyabhama University in Tamilnadu.

In the field of defense applications it will take another 15 years for Indian research scientists to be aware of the usage, development and applications of nano composites. They have to study about the obstacles involved in the incorporation of nano tubes into matrix. They should have clear cut vision and strategies for improvement in properties. They have to get enlightened in optimizing the fabrication of nano tube enabled materials to get advanced mechanical and transport characteristics.

The defense experts aim to get full results from research in nano technology and nano composite materials for UAV to tackle enemies with lower cost of production and better performance with reliability. This new technology research should focus on protection to army men on field and enable them handle with case. As lowering the weight is a special feature of nano material composites, the Indian soldiers carrying weapons in border area, hilly territories and snowy Kashmir borders, prefer their applications. National Institute of Technology in Suratkal, Karnataka, Defense Materials and Stores Research & Development Establishment in Kanpur, Manipal Institute of Technology in Manipal and Sri Manakula Vinayaga Engineering College, Pondicherry, are some of the leading universities, which deal with nano composite research works.

Scientists

Scientists involved in nano composite are spread all over India, in various universities

Dr.Vinita Vishwakarma; an F Grade Scientist has done researches in Environmental Nano technology, Cytogenetics and Plant Pathology. She has 15 years of experience in the field of Bio-fouling, Thin Films Cement and Bio-Corrosion. She has attended 20 conferences, written in 15 journals and travelled abroad for discussions on nano technology.

Mr.D.Ramachandran – Scientist Grade C has specialized in Organic Chemistry, Synthesis and Characterization of Heterocyclic compounds. He has done much works in Cement, Concrete, Bio-fouling, Bio-corrosion, Synthesis and Characterization of Nano particles Optical and Epiflourescense Microscopes. He has attended more than 10 conferences and written papers in 3 journals on Nano technology.

Mr. Arul, a leading Scientist in Grade C is an expert in Medical Nano Technology. He has 5 years of experience in Field Emission Scanning Electron Microscope with EDS mapping, Scanning Transmission Electron Microscope, Correlative Microscopy, SE2 Detector System and Imaging of all kinds of samples, Metals, Polymer-metal Composites, Nano particles, Thin Film Coating and Microbial samples. His papers have been presented in 5 conferences and written technical research articles in 2 leading journals.

K.Viswanathan, Scientist in Grade C has specialized in Nano Sensors, Embedded Systems and VLSI. He has wide knowledge in Nano particles, Thin Film Coating,

Metallographic techniques and obtained doctorate in Embedded Systems. His research articles have been published in 3 journals and presented papers in conferences focused on nano materials.

Dr.Rajesh Kannan, a leading D grade scientist has specialized in Functional Genomics and Molecular medicine. He has wide expertise in the field of Nano particles, Bio Functional Analysis and Small Molecule Drug Discovery. He is engaged in 2 vital projects in Nano bio technological and Medical bio-technology research field utilizing Zebra Fish as model system. Imaging and Nano medical science applications are his field of interest.

His technical research articles have appeared in 10 journals and have been honored in 9 leading international conferences on nano composites.

Dr.J.Brijitha has wide experience in Nano particles Dispersions, Carbon Materials and Soft Condensed Matter. In 10 journals her papers have been printed in black and white and have rendered lectures in 18 conferences.

Dr.T.S.Shyju, a D grade Scientist has specialized in Material Science. She has done dedicated research works in Fabrication of Poly Crystalline and Nano Crystalline Silicon Solar Cells. His speeches in 25 conferences and his research papers in 12 journals have worn laurels all over the world.

Gobi Saravanan, a Scientific Assistant has 3 years of experience in Thin Film Coating and Corrosion Science. In 2 conferences his papers on these topics have been submitted.

G.Pradhaban, another Scientific Assistant has good knowledge in Pulsed Layer Deposition Technique, X-ray Diffraction and Thin Films. He has presented papers in 2 conferences.

Dr.P.Kuppuswami, a Senior Scientist has extensive talents in Nano Science, Nano Technology and Materials Science, Electron Microscopy and Surface Engineering. He has served in Indira Gandhi Centre for Atomic Research (IGCAR) and retired from service as H Grade Scientist. He has presented papers in 140 conferences and his papers on

Nano Technology have found honourable place in 87 journals. He has been awarded 5 patents.

Dr.A.Madankumar, C grade scientist has specialized in Molecular Targets and Cancer Nano Therapeutics. He obtained PhD. in Cancer Biology at Madras University. He has attended 10 conferences to present papers and 7 journals have published his research works.

India needs more and more scientists to create awareness to one and all.

Nano Technology for Agriculture

The nano scientists in India focus more on agricultural production, which is a major, factor to safe guard the families of our wounded Ex Army men and soldiers. Their family members, who do cultivation for their livelihood, warmly welcome these nano innovations for their crops.

Ex army men's dependence on agriculture

Most of the Indian ex-army men in India are in ill-health condition.

Their limbs are amputated, kidney removed and eyesight in worst condition.

Hence to manage for the rest of the life they depend upon agricultural lands offered by the government. To assist their agricultural crop production, nano technology plays a vital role. The scientists focus more on agricultural production, which is a major factor to safeguard the wounded soldiers in India. Their family members, who do cultivation for their livelihood warmly, welcome these nano innovations for their crops.

Some of the latest inventions are as follows:

Effect of Green Synthesized Zinc Nano particles on the Physiological Growth and the Enzyme activities of Groundnut Plants (Pot-Culture)

As per the words of the scientist Sri Sindhuja Kalpa "The Synthesis of Green metal nano particles using leaf extracts is emerging into an exciting and the most significant branch, called Green Synthesis.

Method adopted

Although, the physical and chemical routes are in practice, this specific branch of green synthesis is much more preferred due to the unique characteristics like not-toxic mode environmental- friendly, less labor in handling and easy to perform in the present investigation.

Phytogenic zinc nano particles were synthesized from the leaves extracts of Parthenium Hysterophorous and Senna Auriculata following the green route.

The synthesized zinc nano particles were thoroughly characterized using Ultraviolet-Visible (UV-Vis) Spectrophotometer, Fourier Transfrom Infrared (FTIR) Spectrophotometer, X-ray diffractometer (XRD), Particle Analyzer, Scanning Electron Microscope (SEM), Energy Dispersive X-ray Analysis (EDAX), Transmission Electron Microscope (TEM) and Inductively Coupled Plasma-Optical Emission Spectrophotometer (IPS-OES).

Nano particle size

The absorption edge of zinc nano particles was recorded at about 330nm. The microscopic analysis revealed poly-dispersed, spherical and triangular shaped zinc nano particles with an average particle size of 22nm.

The measured zeta potentials of 80 mv and 117 mv indicated very high dispersion stability of zinc nano particles.

Treatment

The synthesized zinc nano particles were applied to the groundnut (peanut) pot- culture in three different treatments, against the controls and estimated the soil microbial population, soil exo-enzyme activities and the physiological growth parameters of peanut plants in 2 regular time intervals of 30 days and 60 days of sowing period.

Observation

Also the increase in growth parameters from day 30 to day 60 was observed.

Induced growth of soil microbial population, soil enzyme activities and physiological traits of peanut plants was observed for the zinc nano particles applied treatments, when compared to the controls.

Result

Parthenium Hysterophorous leaves extracted zinc nano particles applied treatment showed better results, when compared to Senna Auriculata leaves extracted zinc nano particles applied treatments. Among the three treatments, treatments-1 showed good result, as the zinc nano particles were applied in greater volume for treatment-1 (15ml), compared to treatment-2 (10 ml) and treatment-3 (5ml).

Scientists involved are: Sri Sindhuja Kaipa, O.M Hussain T.N.V.K.V Prasad and Panneerselvam

Address: #1167, I floor, I cross, II Block, BEL Lay out, Vidyaranyapura, Bangalore-560097, Karnataka, India

Phone: 918041743782

Mobile: 919902375005

Email: nano.kssindhura@gmail.com

Inhibitory Activity of Nisin functionalized Gold Nanoconjucate against Food Borne Bacterial Pathogens

Method of operation

Scientist Pradeepa, Nano Scientist from Karnataka says "Food borne diseases encompass a wide spectrum of illness and it is caused mainly by microbial contamination. Nisin is best known Bacteriocin and have been used as food preserving agent for various fermented food stuffs. However, Nisin is an effective inhibitor of Gram-positive bacteria.

but it is not effective against Gram Negative Bacteria. This is due to the presence of complex outer membrane, which acts as a permeability barrier of Nisin action.

Fabrication

In the present study, to enhance the interaction of Nisin and bacterial cell wall, Nisin was fabricated with gold nano particles (GNPs). GNPs were initially prepared by Probiotic Exopolysaccharides. The nano particles functionalized with Nisin were analyzed using UV Visible Spectrophotometer, Zeta Potential, TEM and FT-IR.

Test conducted

Later the nanostructures were tested against various pathogenic bacteria isolated from spoiled food sources. Nisin functionalized GNPs (NGNPs) showed clear inhibition zone against Gram Negative bacteria, whereas Nisin and GNPs alone did not. The MIC and MBC of NGNPs were significantly reduced from 5mg / ml to <1mg /ml. Electron microscopy analysis revealed that NGNPs action was based on membrane destabilization and followed by cell death.

Further, Bio-compatibility findings revealed that, NGNPs unable to lyse human RBC and indicative of non toxicity. These results indicated that, NGNPs can be used as effective antimicrobial agent in food preservation".

Scientist: Pradeepa

Address: Research scholar, Department of Biotechnology N.M.A.M. Institute of Technology, Karkala-574110, Karnataka, India

Phone: 918258281264

Mobile: +919945508574

Email: bio.pradeepgowda@gmail.com

Organization: N.M.A.M. Institute of Technology

Co-Scientists: Udaya Bhat .K, Vidya S.M and Prashant Huligol

Turmeric, naturally available colorimetric receptor for quantitative detection of Fluoride and Iron

Turmeric (Curcuma Longa) an important food ingredient has been used for ancient times for various biological applications but to date it has not been considered for the colorimetric detection of anions and cations.

Significant change

The scientist Pravin Patil says "In this work, the application of turmeric solution for the Colorimetric detection of fluoride (F) and iron (Fe3) ions was demonstrated. Results showed a significant colour change from yellow to blue and yellow to brown upon addition of F ions and e3+ions respectively. The detection mechanism was investigated using curcumin, a major component of Turmeric powder.

The change in colour and fluorescence quenching was attributed to the formation of Receptor complex with Fluoride and Fe3 ions, which resulted in intra-molecular charge transfer transition. The mechanism of binding has been confirmed by UV-Vis, 1H NMR and fluorescence titrations.

Detection kit

The present detection concept was further explored to develop a low cost reusable Fluoride detection kit, which has the ability to detect Fluoride ions at a very low concentration such as 1ppm in both aqueous and organic medium showing its potential to use for real time applications".

Scientist involved is: Pravin Patil

Address: Jain Global Campus Jain University Jakkasandra Post, Kanakapura Taluk,

Ramanagara District Bangalore -562112, Karnataka, India

Mobile: +918123611925

Email: praveen.patil@jainuniversity.ac.in

Organization: Centre for Nano and Material Sciences Jain University

Magesh.P. Bhat, Shriram, G, Uthappa U.T, Madhuprasad and Mahaveer D.Kurkuri are other

Scientists involved in this research.

Synthesis of Soybean Mediated Silver Nano particles and its application against Bacterial

Blight of Pomegranate

Chikkanaswamy, a leading Nano scientist from Dharwad says "Bacterial blight to

Pomegranate caused by Xanthomoans axonopodis PV. Punicae has affected the growers to the extent of 60% of fruit loss.

Usage of Nano technology

Management of this disease using nanotechnology is a novel method and challenging.

In the present investigation, silver nano particles (AgNps) were synthesized from soybean seed extract. Ten per cent of soybean seed extract (10ml) was mixed with 1mM of silver nitrate (50ml) and exposed to bright sunlight and dark condition for 5hrs separately.

Change in Color

Change in color (dark orange) was observed immediately, when exposed to sun light. Further characterization of AgNPs was confirmed by UV –Visible Spectrophotometer, which showed maximum absorbance at 425 nm size (<100nm) and shape (spherical to irregular) of AgNps were characterized by atomic force microscope.

Test

AgNps were tested against X. axonopodis PV Punicae by paper disc technique, on nutrient glucose agar medium.

AgNps showed slight less zone of (9.25mm) inhibition compared to AgNO3 alone (12.50mm) and streptocycline (11.50mm) after 48 hrs of incubation.

Phytotoxicity AgNps was not observed on tomato seedlings at all tested treatments under glass house condition".

Scientist: Chikkanna Swamy

Address: University of Agricultural Science, Dharwad-580005, Karnataka, India

Phone : 918362744321

Mobile: 918762067238

Email:chikka8hbr@gmail.com

Organization: University of Agricultural Science, Dharwad, Karnataka

Co-scientists: Nargund V.B, Mathu S.Giri, Hasansab A.Nadaf and Hulagappa

Characteristic and scope of Biopolymer Encapsulated Nanotised Gypsum Based Nano Reclaimants for Reclaiming Marginal Productive Salt affected soils and water
Affected soil

Nano Scientist Ajay Kumar Bharathwaj from Haryana says "Soil sodicity poses threat to germination of seeds, resistance to roots development and non availability of water to plants for growth due to osmosis. Such adverse conditions make the salt affected soils unviable for productive agriculture.

Besides decreasing productivity of land, these conditions also affect vital land conditions, such as changes in soil biotic forms due to presence of salts, soil erosion due to increased dispersibility, flooding of land due to deteriorated soil structure, low ground water recharge, threatening agro-ecological balance and human and animal health.

Nano technology

Process of reclamation of sodic lands and waters with coarse, poor quality Gypsum is energy intensive requiring high rates of application (10-20 t ha -1 for reclaiming top 30 cm of soil) and high water use (6-12ML ha-1).

Nanotechnology based initiatives in developing reclaimants for salt affected soils have perceivable advantages. Bringing reclaimants, such as gypsum, to nano scale enhanced their reactivity and solubility, which is otherwise only 0.25%.

Encapsulation of Nanotised forms in bio –polymers has two pronged effect wherein active principle exchanges Na and polymers enhance soil conditions by soil particle binding and aggregate stabilization.

Polymeric components

Polymeric components help generate proper soil conditions by increased soil porosity, and help in leaching of exchanged Na salts from soil profile. New modes of application have been conceptualized for easier point specific applications.

The developed nano materials have opened new doors for enhancing reclamation efficiency of salt laden soils and waters".

Scientist involved: Ajaykumar Bharathwaj

Address: ICAR-CSSRI, Kachwa 1 Road, Karnal-132001, Haryana, India.

Phone: +91-9184-2209359

Mobile: +91-9467894326

E mail:ak.bhardwaj@icar.gov.in

Organization: CAR-Central Soil Salinity Research Institute

Green Synthesis of Neem based Silver Nano particles and their efficacy against

Plant Pathogens

As per the popular scientist from Dharwad Dr. V.B.Nargund "Nano technology is an emerging research area in agricultural and allied fields. Present research works mainly focus on plant health management by Green technology.

Nano particles

The Nano scientist from Dharwad Dr. Nargund says, "Synthesis of silver nano particles using Neem leaves extract as reducing agent and silver nitrate as a precursor. Twenty per cent of Neem leaves extract and 1mM of silver nitrate was mixed together in proper combination and kept for continuous stirring under dark condition for 6 hrs.

Preliminary confirmation of formation of AgNPs by change in colour from light yellow to dark brown, further characterization AgNPs was done by UV Visible spectrophotometer and Atomic force Microscope.

Maximum absorbance was observed at 445 nm and particles sizes were less than 100nm.

Test and observations

Efficacies of Neem based silver nano particles were tested against plant pathogens viz., black rot of cabbage caused by Xanthomonas campestris PV. Campestris and banana anthracnose caused by colletotrichum musae.

No anti-bacterial activity of AgNPs was observed against xanthomonas c. PV campestris but antifungal activity of same AgNPs showed 100% spore inhibition against c.musae even at low concentration of 4.5ppm. Further no phytotoxicity was observed on tomato seedling, when tested under glasshouse condition". Poster Presenter: V.B.Nargund

Address: Department of Plant Pathology, University of Agricultural Sciences Dharwad -580005, Karnataka, India

Phone: 918362744321

Mobile: 919448797254

Email: nargund56@gmail.com

Organization: University of Agricultural Sciences, Dharwad, Karnataka India

Co author: Chikkanna Swamy, Madhu, S.Giri, Divya Jagana and Hasansab A. Nadaf

Comparative Anti-microbial Study of Murraya Koenigii Flower Extracts and its Nano particles Formed

The Nano scientist from Bangalore Dr Huda Wakeel says, "Nano particles are being used therapeutically as drug carriers, in which the active principle is adsorbed.

Single step process
Micro-organism and plant extracts, which contain active anti- fungal and anti-microbial principles, may be used as an alternative to chemicals drugs delivered for therapeutic purposes green synthesis route for nano particles is a very spontaneous, economic method, which is non- toxic to the human health and the environment thus its advantage over physical and

chemical means. Nano particles from plant extract using this Green Synthesis Technique is a single step process, suitable for large scale production. This technique has been used to make Silver, Copper and Iron Nano particles from the plant extracts.

Usage of Nano Particle

Antimicrobial study of such nano particles of flower extract Murraya koenigil is done on different clinical bacterial pathogen Escherichia coli, Pseudomonas aeruginosa and Staphylococcus aureus bacillus cereus and some pathogenic fungi Trichophyton simii and Aspergillus Niger.

The usage of nano particles show greater antimicrobial activity as compared to the native extracts.

Silver nano particles show the highest increase in antimicrobial activity especially in case of gram-negative bacteria, where the anti- bacterial activity shows a 100% increase.

Scientist experimented: Huda Wakeel.

Address: 88/1, Gottigere, Bangalore-560083, Karnataka, India

Phone: 918040250520

Mobile: 919008423595

Email: viceprincipal@tjohngroup.com

Organization: T.John College

Co-Scientists: Sharwan Malviya and Nilajana Basu

Current Research Projects in India

Utilization of silica reinforced silicon nano composites coating on aluminum surface will make it anti-corrosive. Such materials will be highly recommended for aerospace and electronic circuits used for Indian defense equipments.

Nano Composites for Defense Applications

The nano technology is paving way for a big turnaround for Indian defense equipments.

Since the neighboring countries are extending their research works in a much faster rate than us, the research on production of nano composites for defense applications is the hot topic now in India.

At present they are breaking their heads about utilization of new chemicals and materials to develop nano composites exclusively for defense applications.

Fibers

Initially very few quantities of tangled nano tubes were produced. This led to creation of non-oriented mats. Later spinning of nano tubes into fibers in a polymer matrix was done. This was used for mechanical and electronic fiber applications. By incorporating nano tubes mats with a polymer, continuous sheets or films of nano composites were produced. These nano tubes will add unique mechanical properties like hardness and strength for the films. They also improve the conductivity of the new material.

Till recently the production of polymer fibers were done for extruding fibers of micrometer sizes. But later on electro spinning method was adopted to get pure polymer and polymer nano composite fibers with 200nm to 300nm diameter. It is important to know that the nano particles were seen in highly aligned state in an electro spun nano composite fibers. The official and mechanical properties needed for defense equipments are achieved by it.

A simple way to make non-woven fabric to attain electrical conductivity is that the non-woven mat will be exposed to a high intensity light source like flash tube. This enabled incorporating fibers at cross over points of contact easily and quickly. To join the fibers in the required pattern a mask is used.

The research works in these materials play a significant role in sensors, microwaves absorption, electrically conductive fabrics, electromagnetic shielding, actuators and capacitors for micro UAVs in Indian defense sector.

Microwave Absorbers

There is a method to mix Polypyrole nano composites having Iron oxides, Titanium oxide, Tungsten oxide and Tin oxide. Magnetic property was achieved by polymerizing Pyrrole having a dispersion of nano particle metal oxides.

By altering the concentration and orientation of nano tubes we can achieve a change in electrical conductivity and dielectric losses.

The quantity of nano tubes to be added is only small to get desired results. The nano scientists have studied to use

CNTs for creating economical microwave absorbers to have a range from 8GHz to 24 GHz. These materials are a great demand for producing condensers integrated into load carrying structures for Unmanned Airborne Vehicle, high strength CNTs Polymer fibers for energy absorption and electromagnetic shielding.

Polymer Optical Fibers

"Polymer optical fibers play a vital role in optical applications like telecommunications and optical computing. They alter the refractive index of the connecting optical fiber. This can be produced in large quantities and cheap with production cost. By incorporating nano particles with many refractive indices to the polymer, we can achieve this", report the scientists.

The following materials are utilized for this purpose and added to Polymethyl Methacrylate to alter the refractive index:
1) Alumina
2) Zirconia
3) Silica

"Resistance to scratching and abrasion of the fiber can be improved by adding ceramic Nano particles", say the Indian stalwarts in nano composites.

Lubricants

It is found that by addition of WS_2 (Tungsten Sulfide) in small quantities to polymers like Epoxy and Polyacetal we can lower the coefficient dry friction between polymer and a steel disc. At the same time fracture toughness was increased.

Foaming properties

The foaming properties of a polymer can be enhanced by adding nano particles like Silica and Carbon dioxide as blowing agent. Another great advantage of adding nano particle is that they will reduce the rate of burning. Even for shock absorbing and acoustic absorbents non-porous Polyurethane is recommended.

For Space craft

Space craft face problems due to dissipation of static charge. Hence we need a material to possess electrical conductivity and also be stable to manage intense ultraviolet radiation, atomic oxygen, rapid changes in temperature and charged particle irradiation. This is possible by a Polyimide Nano composite having 0.03 wt % CNTs.

Styrene-butadiene-styrene clay nano particles nano composite has the required resistance to radiation. The flake like clay particles acts in a passive way to protect the polymer from radiation. Further the nano particle behaves as active link for broken polymer chains, which are on to the nano particle chain.

Resistance to degradation

"Usually polymers are unstable under ultraviolet irradiation and will start degrading after few days. With lowered strength and fracture toughness it will become brittle in due course. This problem is solved by addition of nano particles of Titanium Oxide to Epoxy / Carbon Fiber composite. The mechanical strength is also enhanced by 85 %", say the scientists. By addition of nano particles of Titanium dioxide to Epoxy / Carbon fiber composites we can alter the Epoxy matrix.

This results in resistance to degradation by ultraviolet irradiation.

Fire Retardation

By incorporating flake like clay nano particles it will lower the diffusion of polymer decomposition volatiles to the burning surface and lower diffusion of air into the polymer. Usually polymer has low fire resistance. They will quickly catch fire and give out huge quantity of toxic gases, heat and soot. These problems can be solved by addition of clay nano particles. Polypropylene CNTs Nano Composites also showed improvement. So, this characteristic is vitally used for submarines, aircraft and ships.

Anti-Corrosion

Utilization of silica reinforced silicon nano composites coating on aluminum surface will make it anti-corrosive. Such materials will be highly recommended for aerospace and electronic circuits used for Indian defense equipments.

Camouflage materials

The defense sector needs tunable camouflage materials, which are highly visible or totally concealed presence according to the condition.

The scientists and innovative thinkers have stated that High Contrast Electro Chromatic Nano composite based on Poly Ethyleneimine and Prussian Blue Nano particles will be an ideal choice for defense needs. They remark that fully switchable reflective tri-color space coating is available for defense use as dynamically tunable camouflage material.

Tetra pack

To avoid the damage to food products by carbon dioxide or oxygen we use tetra pack having many layers of aluminum as a good barrier. This eliminates odor and give air tight packing for the food used by Indian army men in remote areas. The army men engaged in border areas far away from metro cites or villages and in areas where no transportation is available, these tetra pack foods will serve him even for months together in preserved condition. This nano composite container is an ideal one to be safe from fire and corrosion. These packets and containers are made out of rubber or elestomers.

Chemical Sensors

Chemical sensors based on SWNT will respond at faster rate to gas molecules like NO_2 or NH_3. The electrical sensors made out of carbon black polymer will operate at high temperature for the needed sensitivity. But SWNT Sensors will be working in greater sensitivity at room temperature itself.

The scientists have found that free standing nano tube polymer composite films can be utilized to produce nano sensors. This will have one conductive channel having an array of aligned CNTs embedded in a matrix. This Dimethyl Siloxane will support to find the real time physical nature of aircraft wing or chassis, while flying in air. These materials will also help to detect toxic gases used by warriors, flammable gases, solvent vapors and touch sensors.

Nano Composite Based Actuators

It is to be noted that nano composite based actuators have the capacity to lower the power needs and linear motion directly. While dispersing a little quantity a CNTs in a Polyurethane Thermoplastic Polymer it is found that the

Nano composite obtained can preserve and release when needed 50 % of more energy than the conventional polymer. CNTs are added by infrared method or by Joule heating method. It is sufficient that lower quantity of CNTs is added compared to Carbon black normally used as additive.

Due to low recovery stress of the present shape memory polymer we have to add inert silicon carbide nano particles to enhance the recovery stress by 50 %. It is done without disturbing the other properties.

A new type of actuator response operated by an electric field in the presence of MWNTs in nematic elestomers polysiloxane was found out. A composite material was produced with embedded and aligned CNTs dielectric anisotropy.

A new technique has also been developed for producing actuating composite materials with polarisable moieties and CNTs.

The defense men prefer these actuators for UAVs.

The Indian research experts opine that even if the interfacial region is only a few nanometers, in a faster manner the entire polymer matrix will possess a different characteristic than the bulk.

If the interfacial area is extended more, then the polymer matrix attitude can be changed at much smaller loadings.

"The processing methods involved to prepare nano composites with controlled particle size, distance, dispersion and interfacial interaction are important and difficult to execute" say the nano composite scientists.

Materials used by the researchers:

Polymeric Nano Composites are:

1. Nano Clay

2. Carbon Nano Tube

3. Nano Fiber

4. Inorganic particle

Hectorite, Montmorillomite and Saponite are the Smectite type layered Silicates for producing nano composites.

Montmorillomite is widely accepted for usage in polymers due to its higher surface area and surface reactivity. The experts say that it is a Hydrous Alumino Silicate Clay mineral with a 2:1 expanding layered crystal structure. Aluminium Octahedron is sandwiched between the layers of Silicon Tetrahedron.

Their current research activities are supported by the Indian government with sufficient funds and imported laboratory equipments.

Main aim is to face threat detection, reduction in maintenance cost and easy to repair.

****

Made in the USA
Coppell, TX
24 March 2025

47521282R00079